T0261420

State, Science and the Skies

RGS-IBG Book Series

Published

State, Science and the Skies: Governmentalities of the British
Atmosphere
Mark Whitehead

Complex Locations: Women's Geographical Work in the UK
1850–1970
Avril Maddrell

Value Chain Struggles: Institutions and Governance in the
Plantation Districts of South India
Jeff Neilson and Bill Pritchard

Queer Visibilities: Space, Identity and Interaction in Cape Town
Andrew Tucker

Arsenic Pollution: A Global Synthesis
Peter Ravenscroft, Hugh Brammer and
Keith Richards

Resistance, Space and Political Identities: The Making of
Counter-Global Networks
David Featherstone

Mental Health and Social Space: Towards Inclusionary
Geographies?
Hester Parr

Climate and Society in Colonial Mexico: A Study in Vulnerability
Georgina H. Endfield

Geochemical Sediments and Landscapes
Edited by David J. Nash and Sue J. McLaren

Driving Spaces: A Cultural-Historical Geography of England's M1
Motorway
Peter Merriman

Badlands of the Republic: Space, Politics and Urban Policy
Mustafa Dikeç

Geomorphology of Upland Peat: Erosion, Form and Landscape
Change
Martin Evans and Jeff Warburton

Spaces of Colonialism: Delhi's Urban Governmentalities
Stephen Legg

People/States/Territories
Rhys Jones

Publics and the City
Kurt Iveson

After the Three Italies: Wealth, Inequality and Industrial
Change
Mick Dunford and Lidia Greco

Putting Workfare in Place
Peter Sunley, Ron Martin and Corinne Nativel

Domicile and Diaspora
Alison Blunt

Geographies and Moralities
Edited by Roger Lee and David M. Smith

Military Geographies
Rachel Woodward

A New Deal for Transport?
Edited by Iain Docherty and Jon Shaw

Geographies of British Modernity
Edited by David Gilbert, David Matless and
Brian Short

Lost Geographies of Power
John Allen

Globalizing South China
Carolyn L. Cartier

Geomorphological Processes and Landscape Change: Britain in
the Last 1000 Years
Edited by David L. Higgitt and
E. Mark Lee

Forthcoming

Aerial Geographies: Mobilities, Subjects, Spaces
Peter Adey

Globalizing Responsibility: The Political Rationalities of Ethical
Consumption
Clive Barnett, Paul Cloke, Nick Clarke and
Alice Malpass

Living Through Decline: Surviving in the Places of the
Post-Industrial Economy
Huw Beynon and Ray Hudson

Swept-Up Lives? Re-envisaging 'the Homeless City'
Paul Cloke, Sarah Johnsen and Jon May

Millionaire Migrants: Trans-Pacific Life Lines
David Ley

In the Nature of Landscape: Cultural Geography on the Norfolk
Broads
David Matless

Transnational Learning: Knowledge, Development and the
North-South Divide
Colin McFarlane

Domesticating Neo-Liberalism: Social Exclusion and Spaces of
Economic Practice in Post Socialism
Adrian Smith, Alison Stenning, Alena
Rochovská and Dariusz Świątek

State, Science and the Skies

Governmentalities of the British Atmosphere

Mark Whitehead

WILEY-BLACKWELL

A John Wiley & Sons, Ltd., Publication

This edition first published 2009
© 2009 Mark Whitehead

Blackwell Publishing was acquired by John Wiley & Sons in February 2007. Blackwell's publishing program has been merged with Wiley's global Scientific, Technical, and Medical business to form Wiley-Blackwell.

Registered Office
John Wiley & Sons Ltd, The Atrium, Southern Gate, Chichester, West Sussex, PO19 8SQ, United Kingdom

Editorial Offices
350 Main Street, Malden, MA 02148-5020, USA
9600 Garsington Road, Oxford, OX4 2DQ, UK
The Atrium, Southern Gate, Chichester, West Sussex, PO19 8SQ, UK

For details of our global editorial offices, for customer services, and for information about how to apply for permission to reuse the copyright material in this book please see our website at www.wiley.com/wiley-blackwell.

The right of Mark Whitehead to be identified as the author of this work has been asserted in accordance with the Copyright, Designs and Patents Act 1988.

All rights reserved. No part of this publication may be reproduced, stored in a retrieval system, or transmitted, in any form or by any means, electronic, mechanical, photocopying, recording or otherwise, except as permitted by the UK Copyright, Designs and Patents Act 1988, without the prior permission of the publisher.

Wiley also publishes its books in a variety of electronic formats. Some content that appears in print may not be available in electronic books.

Designations used by companies to distinguish their products are often claimed as trademarks. All brand names and product names used in this book are trade names, service marks, trademarks or registered trademarks of their respective owners. The publisher is not associated with any product or vendor mentioned in this book. This publication is designed to provide accurate and authoritative information in regard to the subject matter covered. It is sold on the understanding that the publisher is not engaged in rendering professional services. If professional advice or other expert assistance is required, the services of a competent professional should be sought.

Library of Congress Cataloging-in-Publication Data

Whitehead, Mark, 1975–
    State, science, and the skies : governmentalities of the British atmosphere / Mark Whitehead.
       p.   cm. – (RGS-IBG book series)
    Includes bibliographical references and index.
    ISBN 978-1-4051-9174-6 (hardcover : alk. paper) – ISBN 978-1-4051-9173-9 (pbk. : alk. paper)
       1. Air–Pollution–Government policy–Great Britain. 2. Air quality–Government policy–Great Britain. 3. Science and state–Great Britain. I. Title.
    TD883.7.G7W45 2009
    363.739′2560941–dc22
                                            2009004219

A catalogue record for this book is available from the British Library.

Set in 10/12pt Plantin by SPi Publisher Services, Pondicherry, India
Printed and bound in Malaysia by Vivar Printing Sdn Bhd

1   2009

For Anwen Mair Whitehead (born 12th November 2008)

Contents

Figures and Tables

TABLES

Series Editors' Preface

The RGS-IBG Book Series only publishes work of the highest international standing. Its emphasis is on distinctive new developments in human and physical geography, although it is also open to contributions from cognate disciplines whose interests overlap with those of geographers. The Series places strong emphasis on theoretically informed and empirically strong texts. Reflecting the vibrant and diverse theoretical and empirical agendas that characterize the contemporary discipline, contributions are expected to inform, challenge and stimulate the reader. Overall, the RGS-IBG Book Series seeks to promote scholarly publications that leave an intellectual mark and change the way readers think about particular issues, methods or theories.

For details on how to submit a proposal please visit:
www.rgsbookseries.com

Kevin Ward
University of Manchester, UK

Joanna Bullard
Loughborough University, UK

RGS-IBG Book Series Editors

Preface

My interest in atmospheric government was kindled in 2003 when I came across the British government's *Air Quality Archive*. The *Air Quality Archive* is partly overseen by the Department for the Environment, Food and Rural Affairs (Defra) and provides an online, real-time record of air pollution for urban and rural Britain. Three things in particular struck me about this fascinating fragment of cyberspace. First, was the very notion of an *archive* of the air. Throughout much of modern history, archives, of many different kinds, have provided key ordering devices for scientists and bureaucrats intent on recording different aspects of the natural and social worlds. Yet there was something incongruous about the notion of an air archive: juxtaposing as it does the rigid ordering technologies of the modern world with that most dynamic, fleeting and perpetually mobile of compounds. Was it an attempt to order an unorderable? The second aspect of this air archive was that it was based upon approximately four and half million readings of the air every year. I subsequently discovered that this plethora of atmospheric measurements was the product of thousands of sampling devices and stations that have been established throughout Britain since the early decades of the twentieth century. This phenomenal level of air surveillance appeared to me to mirror an interesting expression of what Michel Foucault has described as *analytical responsibility*. The untiring work that fed the archive seemed to reflect an analytical responsibility towards atmospheric affairs on the part of the British State. I was left wondering where this level of analytical responsibility had come from, and why it was seen as a duty of the State. Third, and finally, my discovery of this atmospheric archive left me wondering what the implications of such an activity could be for my own and others' atmospheric conduct. Until this point I had remained blissfully ignorant of such an extensive record of air pollution, but now I found my transcendental indifference towards daily fluctuations in air pollution levels shattered. I had the power to know the quality of the air I would be breathing in different locations, and the precise composition of the chemical cocktails that it contained. I felt compelled to consider how my own actions were contributing to the mesmerising complexity of air pollution in Britain.

Soon after my discovery of this digital record of the air I began the slow, but always fascinating, study of this governmental archive. I soon realised that the tale behind this archive was a long one, stretching back over 150 years. I also discovered that this was a story that involved the mixing of science and government throughout a variety of geographical locations in Britain. The research required to develop this story has taken me to the municipal records of large urban corporations such as London, Birmingham and Glasgow; the archives of a range of national government departments including the Meteorological Office, Department of Scientific and Industrial Research; the Department of the Environment; and the records of various scientific laboratories. This research project also involved studying the personal records of coal officers, nuisance inspectors, inventors, medical officers, doctors, police officers and smoke abatement societies, who had all assiduously contributed to the formation of a record of British air pollution.

Three intellectual traditions have consistently helped me to interpret the varied and voluminous nature of available archival material on the relationship between science, government and air pollution in Britain. First, the writings of Michel Foucault on the history of government, and the relationships between government and personal forms of conduct, have provided a rich methodological and theoretical terrain for my work. Second, the collective writings of scholars analysing different aspects of the sociology of scientific knowledge (including Vladimir Jankovic, Bruno Latour and Steven Shapin) have enabled me to understand better the connections between the State and science. Third, and finally, I have been strongly influenced by the work of contemporary geographers (including Stuart Elden, Matthew Hannah and Simon Naylor) who have exposed the critical role of space, as well as history, in shaping the construction of various forms of socio-environmental knowledge. Collectively these interconnected intellectual traditions have enabled me to understand how and why the production of atmospheric knowledge occurs at the creative intersection between government, science and space.

The story that follows will, I think, be of interest to scholars working in geography, the history of science, science and technology studies, the political sciences, and Foucault studies. I also hope that this book will have relevance for all those concerned with the political and scientific processes that shape what we know about the changing contents of the atmosphere and structure our varied relationships with the air.

Acknowledgements

During the completion of this research I have become indebted to the help of various individuals and institutions. I would like to acknowledge the support of the Department of Geography at University College London, and in particular Richard Munton, who kindly provided office space and valuable advice during the completion of early archival research. I would also like to thank the University of Wales, Aberystwyth Research Fund, which provided financial support for my research activities. I am also indebted to staff at the Statistical Division of the Department for the Environment, Food and Rural Affairs, and the Carbon Trust, who gave up their time to speak with me about atmospheric government. I would also like to thank Richard and Stephanie Rugg and Tim and Carol Cresswell for providing me with regular overnight accommodation and wonderful hospitality during my numerous trips to London. Thanks also go to Mark Goodwin and Ann Barlow for the use of their spare room on my trips to the Meteorological Office Archives in Exeter, and Jon Oldfield for facilitating a very useful research trip to Glasgow. Finally, I would like to thank the tireless efforts of staff working at Birmingham City Archives, the British Library (London), London Metropolitan Archives (Farringdon), the Meteorological Office Archives (Exeter), the Mitchell Library (Glasgow), the National Archives (Kew), the National Library of Wales (Aberystwyth), and the Royal Society Library (London), in supporting my efforts to track down the material upon which this volume is based.

Beyond the research that has supported this project, I have also been very fortunate to receive a wide range of help in the writing of this volume. I am indebted to the Arts and Humanities Research Council (Grant number AH/F003056/1) for funding the period of research leave during which large portions of this book have been written. Sections of this volume were written in a wide range of different locations. I would thus like to acknowledge the supportive and creative environments for writing provided by the Centre for Alternative Technology (Gwynedd), The City Institute (York University, Toronto), Mary Immaculate College (Limerick), the Combined Universities of Cornwall (Penryn) and the National Library of Wales. The majority of this volume has, however, been written in two locations: in my home village of Cnwch Coch, overlooking the hills of the Magwr Valley; and in my office in the Institute of Geography and Earth Sciences (IGES, Aberystwyth University). To this end I would like to thank, respectively, my wife Sarah, and colleagues in IGES for making both locations such happy places to think and work. This book has also benefited from enormous levels of intellectual support, advice and guidance. I would like to particularly thank Simon Naylor for his consistent interest in my atmospheric work and for our late night discussions of all things geographical. My appreciation also goes out to Rhys Jones for our daily discussions on state theory and writing *inter alia*; Gareth Hoskins for his advice on museum studies and the political nature of exhibition space; Peter Merriman for his valuable guidance on archival research; and Matthew Hannah for some wonderful insights into the work of Michel Foucault. The comments offered by seminar audiences at the Geography Division of the Open University; Department of Geography at the University of Exeter; and the University of Exeter in Cornwall have also been enormously helpful in the completion of this book. I am especially indebted to contributors to the special conference sessions 'Spatial Technologies/Technological Spaces' and 'Atmospheric Geographies' convened at the Annual International Conferences of the Royal Geographical Society (with the Institute of British Geographers) in 2005 and 2007, respectively. The following people have also provided valuable insights and advice: Anna Bullen; Luke Desforges; Deborah Dixon; Kate Edwards; Ian Gulley; Roger Keil; Vladimir Jankovic; Martin Jones; Kelvin Mason; Robert Mayhew; Richard Noakes; Carol Richards; James Ryan; Heidi Scott; Ruth Stevenson; Marc Welsh; Mike Woods; and Sophie Wynne-Jones. Finally I would like to acknowledge members of the Editorial Board of the RGS-IBG Book Series and three anonymous reviewers for their comments on the manuscript; Kevin Ward for his great enthusiasm and support for this project; and Jacqueline Scott for her valued advice during the final production stages of this manuscript.

Abbreviations

ACAP	Advisory Committee on Atmospheric Pollution [Meteorological Office]
ADMN	Acid Deposition Monitoring Network
AURN	Automated Urban and Rural Network
AUN	Automated Urban Network
B.Cit.Arch	Birmingham City Archives
CEH	Centre for Ecology and Hydrology
CIAP	Committee for the Investigation of Atmospheric Pollution
CUEP	Central Unit on Environmental Pollution
Defra	Department for the Environment, Food and Rural Affairs
DoE	Department of the Environment
DSIR	Department of Scientific and Industrial Research
ECDIN	European Chemical Data Information Network
ECEP	European Community Environment Programme
EMPACT	Environmental Monitoring for Public Access and Community Tracking (EPA, USA)
G.Cit.Arch	Glasgow City Archives
HC.PP	House of Commons Parliamentary Papers
ICAPR	Interdepartmental Committee on Air Pollution Research
IPCC	Intergovernmental Panel on Climate Change
IRPTC	International Register of Potentially Toxic Chemicals
L.Met.Arch	London Metropolitan Archives
LRTAP	Convention on Long-Range Transboundary Air Pollution (Europe)
M.Off.Arch	Meteorological Office Archives
NAEI	National Atmospheric Emissions Inventory
NAPS	National Air Pollution Survey
NEGTP	National Expert Group on Transboundary Pollution

NERC	Natural Environment Research Council
NETCEN	National Environmental Technology Centre
N-MVOC	Non-Methane Volatile Organic Compounds
OUP	Ozone Umbrella Programme
OECD	Organisation for Economic Cooperation and Development
RCEP	Royal Commission on Environmental Pollution
RGAR	Review Group on Acid Rain
SCCB	Standing Conference of Cooperating Bodies
SDMN	Sulphur Dioxide Monitoring Network
SSK	Sociology of Scientific Knowledge
TNA	The National Archives (Kew, London)
UKNFC	UK National Focal Centre for Critical Loads Modelling and Mapping
UNECE	United Nations Economic Commission for Europe
UNEP	United Nations Environment Programme
VOC	Volatile Organic Compounds
WSL	Warren Spring Laboratory

Chapter One

Introduction: Space, History and the Governing of Air Pollution

On 700 Years of Air Government

It is the year 1307 in medieval London. Rumours are abound that one denizen of the fledgling metropolis has been subjected to a gruesome penalty for perpetrating the most novel of crimes. This unnamed individual, it was claimed, had broken the recent Royal Proclamation banning the burning of sea-coal in the city. The punishment meted out to this early atmospheric felon, or so the tale goes, was torture, hanging and ultimate decapitation![1] While it seems unlikely that such a punishment was ever actually carried out,[2] it was perhaps the nature of the crime, as much as the extreme form of the purported penalty, which would have concerned the fourteenth-century urban dweller. Before the Royal Proclamation of 1306 the idea that polluting the air could be deemed a criminal offence was simply inconceivable. The age of British atmospheric government had begun.

It is the year 2007 in post-industrial Britain. According to latest government figures, over four million readings have been made of the British atmosphere this year from a network of over 1500 government-sponsored air pollution monitoring stations.[3] This never-ending process of 24-hour air surveillance has recorded the varying concentrations of a heady chemical concoction of pollutants including ammonia, sulphur dioxide, trace metals, oxides of nitrogen, organic micro pollutants, particulate matter and various hydrocarbons *inter alia*. Only a small fraction of the incomprehensible volume of atmospheric knowledge produced by the British government in 2007 will be used to support the prosecution of air polluters. None has been utilised as a basis for summary execution!

This book explores the history of contemporary systems of air pollution government in Britain. To this end it is, in part at least, interested in what has been happening in British atmospheric government between 1307 and 2007.

It is clear that this conveniently demarcated historical epoch has been characterised by some profound changes in the ways that political authorities organise the governance of air pollution. It is also evident that a detailed study spanning such a long historical reach would be beyond the scope of a single volume. Consequently, while broadly positioned within this 700-year era, analysis is primarily concerned with the systems of air pollution government that have emerged in Britain since 1843. The year 1843 is significant in the history of British air pollution government for two primary reasons. First, it was in this year that the Parliamentary *Select Committee on Smoke Prevention* was established in order to discuss the ensuing problems of atmospheric pollution in industrial Britain.[4] Second, and as part of the operation of this Committee, 1843 witnessed the first systematic attempt made by the British government to forge close working relations with scientists in the crusade against the contamination of the air. More will be said of the 1843 Select Committee in Chapter Two, but at this point it is important to see this Committee as a crucial historical moment in the emergence of the knowledge-intensive and scientifically grounded systems of air pollution government that are now commonplace in Britain. It was the beginning of what this volume refers to as a system of atmospheric *government with science*.

The notion of atmospheric pollution is a complex and ever-changing category of analysis that has, at different times, incorporated germs, disease, dust, pollen, grit, smoke, fog, soot, sulphur dioxide, lead, radioactive materials, pesticides, chlorofluorocarbons, carbon dioxide and other visible and invisible substances (see DuPuis, 2004: 1–11). As Thorsheim observes, however, the processes that transform these various substances into pollution occur at the complex intersection between culture and nature (2006: 155; see also Douglas, 1966). Atmospheric pollution involves more than anthropogenic or environmentally produced contaminates simply entering the air. In order to become pollution, contaminates have to work with the pressure dynamics, weather patterns, thermodynamic systems and chemical exchange functions of the atmosphere, and produce culturally, biologically and politically unacceptable/intolerable air conditions. It is for these reasons that analysis will engage with the activities of meteorologists, climatologists, ecologists, chemists, medical experts, civic activists and policymakers who collectively constitute the hybrid science that frames air pollution government in Britain.

Unpacking the Politics of Air Pollution Science and Government

In many ways concern over the axis between atmospheric knowledge and systems of air government has never seemed more important. As I wrote this book global media coverage of the quality of one city's air, and the systems of atmospheric government that are being deployed to combat associated forms of air pollution, have become almost obsessive. In August 2008 Beijing hosted the 29th Olympic Games, but alongside debates around human rights it is miniscule airborne particles (or particulate matter) that caught all of the headlines. These tiny particles were an object of concern for a phalanx of scientists and bureaucrats incorporating the International Olympic Committee, the Chinese government, the World Health Organization, the Beijing Municipal Environmental Protection Bureau and even the World Bank. Beyond the atmospheric hysteria that engulfed Beijing, there are three fundamental issues associated with the city's air quality debate that have direct relevance for the objectives of this volume. First, how and where is the quality of the air measured? Second, on what basis are standards for socially and ecological permissible levels of air pollution determined? Third, how can persistent forms of air pollution be effectively governed? The first question, concerning the scientific practices and locations of air pollution monitoring, was of particular significance in Beijing. In the build-up to the Olympic Games the regular air pollution readings taken by the Beijing Municipal Environmental Protection Bureau were joined by a host of other formal and informal monitoring devices operated by media outlets, international organisations and concerned athletes.[5] With so many different measurement devices, operating in so many different locations, and at various times of day, it was little surprise that there was so much uncertainty concerning the actual levels of air pollution in the city. In terms of setting permissible thresholds for pollution, the World Health Organisation recommended that levels of atmospheric particulates should not exceed 50 micrograms/cubic metre (World Health Organization, 2005).[6] Estimates of particulate air pollution in Beijing, made in the months before the Games, suggested that levels were in excess of 130 micrograms/cubic metre.[7] While providing useful governmental targets for air pollution abatement, as we move through this volume we will see that such thresholds of permissible atmospheric pollution are not always reliable predictors of the potential health (or environmental impacts) of pollution, and are themselves subject to much scientific deliberation. Perhaps the most significant implication of the events in Beijing for this study is the style of governmental intervention that has emerged in response to the identification of harmful air pollution levels. In answer to the air pollution problems of the city the Chinese government took the rather unusual step of closing down polluting factories and plants and, in the event of particularly severe air pollution incidents, removing up to 90% of the traffic from Beijing's roads (see Bristow, 2008).

There are important parallels between the science and government of air pollution in Beijing and the current situation in Britain. While it is important to acknowledge that the levels of air pollution in Britain, and the associated threats posed to the environmental health of its citizens, are not as severe as the current situation in Beijing, air pollution remains a significant governmental issue. Severe air pollution events such as the London smog of 1991 (when nitrogen dioxide concentrations reached their highest recorded levels in Britain) were associated with a 10% increase in the death rate in and around the metropolis (Brown, 1994). A recent report by the Royal Commission on Environmental Pollution estimates that air pollution is, on average, responsible for 24,000 premature deaths in Britain each year, and claimed that the British State had been unsuccessful in addressing increasing levels of chemical pollutants in the atmosphere (Royal Commission on Environmental Pollution, 2007: 35–40).[8] The report also revealed that in 2005 the costs of air pollution to the British economy (in relation to the provision of medical care and lost working hours) were in excess of £9.1 billion (ibid.: 35).[9] As with the situation in China, there remains significant debate in Britain concerning what permissible levels of pollution are, how air pollution should be measured, and the role the government should take on issues of atmospheric pollution. Despite these parallels, however, a clear distinction does exist between the control and monitoring of air pollution in Britain and China: namely the styles of government deployed to address socioeconomic relations with the atmosphere. While China has been able to deploy relatively authoritarian systems of air pollution control in the short term, Britain has witnessed the emergence of very different strategies of air government that reflect a more liberal political tradition. The particular systems of atmospheric government deployed within liberal (and neo-liberal) societies, and the specific mixing of air and social power they involve, constitute a key object of enquiry within this volume.

Conceptual Parameters: Spatial Histories and Atmospheric Geographies

The development of an historical perspective on the government of air pollution in Britain is important because it helps to reveal the contingent political decisions and scientific struggles that have contributed to the establishment of a contemporary apparatus of atmospheric knowledge gathering. History, in this context, helps to assert that what we know about air pollution, and the ways in which atmosphere are governed, are not inevitable parts of closed systems of air science and government, but are legitimate objects of political contestation and potential transformation. Yet the historical perspective developed through this volume does not only seek to position air pollution government in relation to the ways it has changed and evolved through time, but also explores the material conditions under which it has even been possible to conceive of knowing and governing something as large and complex as the atmosphere. In this context, this volume presents a spatial history of air pollution government (see Elden, 2001; Rose, 2007). The notion of spatial history is utilised to reveal that not only have the axes connecting British atmospheric knowledge and government changed over time, but that geography has played a crucial role in the constitution of air government and in shaping the production of atmospheric knowledge. This is an account of history within which space is neither 'static', nor merely a 'cross-section through time': it is rather a '[s]phere in which distinct stories coexist, meet up, affect each other, come into conflict or cooperate' (Massey, 1999: 274). Two conceptual frameworks support the spatial history developed in this volume. The first is the history of governmental reason (or *governmentality*) developed by the French philosopher Michel Foucault (see 2007 [2004]; see also Dean, 1999). Foucault's governmental histories are important because they focus explicitly on the connections between knowledge and power within liberal societies, while revealing the historical specificities of governmental modes of rationality. The second conceptual framework that informs this project is a programme of research that is known collectively as the *Sociology of Scientific Knowledge* (see Shapin, 1995). This broad body of scholarship incorporates work within the history of science and science and technology studies, and collectively draws attention to the conditions under which scientific knowledge is produced and the processes in and through which such knowledge forms circulate. While more will be said of the connections and tension between these two intellectual traditions in Chapter 2, I contend that both provide crucial contexts for the development of an integrated spatial history of air science and government pursued within this volume.

While focusing specifically on the spatial and historical dynamics of air pollution government in Britain, this volume is also indebted to a much broader intellectual re-engagement with atmospheric questions within the discipline of geography. The commitment of the geographical discipline to the development of holistic scientific perspectives on the earth's environmental systems has meant it has had a long dedication to the study of the atmosphere as a complex socio-environmental system. It is in this context that geography has long provided a supportive home to climatologists, meteorologists and atmospheric scientists of various kinds. In recent years, however, there has been a distinct increase in work by so-called 'human geographers' addressing various aspects of atmospheric study. These new atmospheric pioneers are utilising the perspectives provided by anthropology, economics, the social sciences and history in order to develop new analytical perspectives on the air. Recent work by geographers has consequently explored the economic commodification of the atmosphere (Randalls & Thorne, 2007); the historical geographies of meteorological knowledge production (Naylor, 2006); the associations between art and the representation of air pollution (Thornes, forthcoming); and the complex relationships that exist between the climate and human history (Endfield, 2007, 2008). Crucially, and in keeping within the intentions of this volume, the development of these new atmospheric geographies has not been based upon an antagonistic relationship with the physical sciences of the atmosphere (or the establishment of an aerial front, if you like, in the science wars), but on creative dialogues and new research partnerships between human and physical geographers.[10] Through a consideration of the spatial governance of air pollution, this volume hopes to contribute to this synthetic science of atmospheric study: a scientific project that embodies the integrative ethos of the geographical discipline as a whole (see Massey, 1999).

Timeframes and Conceptual Enclosures: On the Structure of the Book

Although the issues of air pollution government, science and knowledge production weave their way throughout the different chapters of this volume, the book has been structured in order to facilitate detailed considerations of both different historical time periods and key conceptual questions. The organisation of this book has thus been deliberately designed in order to make the volume both comprehensive (in terms of the preservation of an historical narrative on modern systems of air pollution government in Britain) and comprehensible (in relation to the ways in which individual chapters conceptually interpret key themes in the history of British air government). What results is a series of chapters that simultaneously contribute to an overall historical infrastructure – revealing the development of air pollution science and government – while also facilitating a more detailed conceptual analysis of the key issues that have characterised modern atmospheric government in one State. It is in this context that the empirical chapters of this volume constitute interlocking, but not sequential, histories of air pollution government and science. To a certain extent the way in which any book is divided is an arbitrary exercise of ordering on behalf of the author. However, in order to be consistent with the historical methodology I establish within this volume, I have attempted to ensure that, while different chapters facilitate certain forms of conceptual focus and analysis, they reflect evolving historical processes of atmospheric government as opposed to an adaptation of history to suit preconceived theoretical concerns.

Chapter Two begins the historical narrative that structures this whole volume by reflecting on the 1843 Parliamentary Select Committee on Smoke Prevention. The majority of this chapter is, however, devoted to charting the key conceptual concerns of this volume. It outlines the key conceptual and methodological contours of Michel Foucault's history of government and work within the sociology of scientific knowledge, while explaining the value of combining the insights of these two intellectual traditions within the study of air pollution science and government. Chapter Three constitutes the first main empirical chapter of this volume. It explores the origins of modern forms of air pollution government within various urban nuisance prevention and sanitary authorities and focuses on the particular challenges facing the creation of an optical regime of air science and government. Chapter Four moves on to consider the role of clean air exhibitions and associated educational initiatives in enabling emerging systems of scientific knowledge concerning the production and extent of air pollution to become referential contexts for personal systems of atmospheric reform and self-government. In the wake of the first International Smoke Abatement Conference, which was held in London in 1912, Chapter Five considers a series of attempts that were made to form the first instrument-based studies of British air pollution. Focusing on the innovative work of key scientists, such as John Switzer Owens and Sir Napier Shaw, and the Committee for the Investigation of Atmospheric Pollution, this chapter analyses the role of technological devices in the constitution of networks of government and scientific networks, and the impacts of the demands for governmental knowledge on the design and implementation of instrumental sciences of air pollution.

Building on the account of early, but highly fragmented, networks of air monitoring instruments, Chapter Six describes the process in and through which a national system of air surveillance was gradually instituted in Britain. Focusing on the development and implementation of the National Air Pollution Survey (that ran from 1961 to 1971) this chapter considers the role of spatial calibration in ordering the study and government of the atmosphere. Chapter Seven describes how the emergence of automated and digital systems of air pollution monitoring transformed atmospheric knowledge production and government during the 1970s and 1980s in Britain. Drawing on notions of telemetric territoriality and digital beings, this chapter explores the impacts of real-time and online atmospheric knowledge systems, and associate simulations of air pollution, on contemporary practices of atmospheric government. In Chapter Eight attention is given to the impacts that new systems of environmental thought and ecological science have had upon the constitution of British air pollution government. Critically questioning the extent to which air pollution government has moved from a concern with human health to an ecological rationality of atmospheric government, it outlines the application of ecologically inspired techniques of pollution analysis throughout different locations in Britain. The concluding chapter provides a review of the key analytical themes that run through the constituent chapters of this volume. In addition to reflecting on key themes, however, Chapter Nine also considers the lessons that a spatial history of air pollution government with science in Britain can provide for the systems of air government that are emerging in response to contemporary forms of climate change and associated atmospheric threats.

In his foreword to the National Smoke Abatement Society's *Smoke Abatement Exhibition Handbook and Guide* of 1936, the then British Minister for Health, Sir Kingsley Wood MP reflected,

> Provision is being made more and more to secure pure water, pure milk, and pure food. But every day we breathe a quantity of air much greater in weight than the quantity of food and drink which we consume (National Smoke Abatement Society, 1936: i).

It is clear that there is no more important, immediate or ongoing challenge to the efficacy of government than the ability to know and regulate the constituents of the air we breathe. The remainder of this volume explores the spatial narratives and entangled scientific endeavours that constitute one State's quest to address this challenge.

Chapter Two

Historical Geographies of Science and Government: Exploring the Apparatus of Atmospheric Knowledge Acquisition

'Men of Science' and the Genesis of British Atmospheric Government

In 1843 the Reverend J.E.N. Molesworth, Vicar of Rochdale and Chair of the *Manchester Association for the Prevention of Smoke*, submitted a petition to Parliament in which he called for a governmental enquiry into the smoke pollution issue (see Ashby & Anderson, 1981: 7). Molesworth's advocacy of smoke abatement, at both a national and municipal level, was marked by a dual belief system. At one level, his commitment to the cause of air pollution control was a moral crusade of social care that was clearly conditioned by his religious beliefs (Mosley, 2001: 119). It was, however, also based upon a firm commitment to the crucial role that 'men of science' would play in solving the smoke problem.[1] Molesworth's petition led to the formation, later that year, of a Parliamentary *Select Committee on Smoke Prevention* with W.A. Mackinnon MP (Molesworth's brother-in-law) as Chair.[2] Although the 1843 Select Committee was neither the first, nor the last, Parliamentary committee to be convened to discuss matters of air pollution, it is significant for two reasons. First, this Committee reflects the culmination of a long and arduous Parliamentary struggle to establish air pollution as a legitimate area of governmental responsibility.[3] Second, as the *Report of the Select Committee* indicates, it sought to unite government officials and men of science in its quest for cleaner air,

In their endeavours to investigate the subject, Your Committee have deemed it expedient to call before them a variety of persons. They have received the evidence of the most eminent men in the science of Chemistry, of Practical Engineers of high reputation, of leading Master Manufacturers and Proprietors of Steam-engines, and of ingenious persons who have devised means and taken out Patents for the Prevention of Smoke.[4]

While the presence of scientific experts on a government committee may appear routine in the context of contemporary relationships between British government and science, it is important to point out the unusual nature of this union of the State and science. While the State had supported a limited number of, so-called, scientific institutions for some time (notably the Royal Observatory at Greenwich, since 1675, and the Geological Survey) prior to the nineteenth century, Rose and Rose describe a 'continuous governmental indifference to science' in Britain (1971: 17–21). In addition to the perceived threat of science to governmental power, this indifference was based upon uncertainty concerning the political and economic value of science. Furthermore, scientists of eighteenth- and early-nineteenth-century Britain struggled to gain recognised professional accreditation and a foothold in the university system (ibid.). It is against this backdrop that the Parliamentary turn to science in the fight against air pollution was so significant.

It is not that the 1843 Parliamentary Select Committee marked the beginning of British governmental concern or action towards matters of air pollution. Following the issuing of the Royal Proclamation banning the burning of sea-coal in 1306 (and the subsequent sovereign actions of King Richard III and Henry V to control air pollution) a series of local initiatives emerged in Britain to regulate the quality of the air. Concerns over the localised odours and smells that were generated from tanneries, brewers, butchers and poor drainage and sanitation systems (what Mieck (1990) collectively refers to as incidences of *pollution artisanale*) were addressed and resolved through the courts leet systems (see Brimblecombe, 2004: 16). Established through Royal Franchise, courts leet were responsible for prosecuting small offences in locally designated territories.[5] Beyond such local legalistic systems of air government there was little resembling the development of a comprehensive governmental strategy for air pollution government before the nineteenth century. Perhaps the nearest Britain came to a national system or air pollution government in the pre-industrial era occurred in the wake of the publication of John Evelyn's 1661 *Fumifugium; or the inconvenience of the aer and smoak of London dissipated.*[6] Evelyn's famous account of air pollution infesting the Royal Palace of Whitehall, and in so doing threatening the health of the monarch and disrupting a crucial seat of government, is a powerful metaphor for the political entanglements between air pollution and the British State that occupies this volume.[7] It was Evelyn's comprehensive vision for tackling the problems of air pollution, and improving the qualities of the metropolis' air, that won him favour with King Charles II and led to him being invited to submit a Bill addressing the air pollution in London to Parliament (Brimblecombe, 1987: 50).[8] For reasons that remain unclear, Evelyn's Bill never passed through Parliament and even after the Great Fire of London his plans for urban reform were not implemented.[9] Nevertheless, Evelyn's call for comprehensive government action on air pollution was clearly a precursor to the expanding role of the British State in atmospheric relations in the coming centuries. Consequently the 1843 Parliamentary Select Committee represents just one moment within emerging governmental strategies towards air pollution.

Among the *eminent men* and *ingenious persons* who gave evidence before the 1843 Committee was Michael Faraday, the British scientist made famous for his discovery of electromagnetic induction.[10] It is significant that Faraday already had links with British government officials, working as he did out of the laboratory of the Royal Institute (the only State-sponsored laboratory in existence at the time) (Rose & Rose, 1971: 17–21). Faraday's deposition before the Committee is indicative of the importance placed by the group on forging stronger links between science and government in the quest for cleaner air. While strongly in favour of the application of government intervention in atmospheric affairs, Faraday cautioned that, '[m]y impression is that, in the present state of things it would be tyrannical to determine that that must be done which at present we do not know can be done.'[11] If government invention within air pollution was to proceed, it appeared that science would have to play a leading role in constructing the systems of knowledge upon which efficacious government action would be based.

In hearing key evidence from 22 prominent scientists and engineers, concerning various aspects of air pollution, the Committee established a strong dialogue between the British State and scientific communities on matters of atmospheric pollution. This dialogue continued with the appointment of two scientific advisors to government on matters of smoke pollution. The two experts in question were Sir Henry Thomas De la Beche (Director of the Geological Society and a great advocate of State support for science) and Dr Lyon Playfair (Professor of Chemistry at the Royal Manchester Institute and later Chemist to the Geological Survey) (see Ashby & Anderson, 1991: 11). The choice of Playfair to act as scientific advisor to the British government is particularly significant given that he was instrumental in making the Royal Manchester Institute one of the leading centres for studying the analytical chemistry of air pollution in the nineteenth century.[12] In 1846, under instructions from the Earl of Lincoln, De la Beche and Playfair produced a report of their studies, *The Means of obviating the Evils arising from the SMOKE occasioned by Factories and other Works situated in large Towns.*[13] In this report De la Beche and Playfair came to the conclusion that while the technologies and techniques existed to effectively abate smoke, the smoke problem persisted because of a lack of effective scientific support for the implementation of local laws and prohibitions.[14] A significant amount of their report was thus dedicated to the problems of scientifically observing and classifying atmospheric effluvia and to the early attempts made by police authorities to provide observational registers of air pollution (for more on this theme see Chapter Three).[15] According to De la Beche and Playfair, if air pollution was going to be effectively governed the input of scientists would be needed in order to provide the indisputable data upon which legal prosecutions could be pursued.

The 1843 Select Committee and the De la Beche and Playfair report did not result in immediate or comprehensive legislative action on air pollution being taken in Britain (it was not until the 1848 Public Health Act that we see the formation of a national legislative clause on smoke pollution, while the first Alkali Act was not passed until 1863). What this period did, however, represent was an enmeshing of science and the State within the tentative construction of the atmosphere as an object of rational governmental reflection and action. In deploying the term government here I am not simply utilising it as a convenient synonym for the State. According to Foucault there is an *irreducible specificity* to the notion of government as a form of power that sets it apart from notions of sovereignty, discipline and political bureaucracy (2007 [2004]: 245). I thus understand government to refer to a form of political power and practice that started to emerge during the seventeenth and eighteenth centuries and combined an ethics of wise care (pastoralism) with an apparatus of knowledge production (science) that was capable of guiding this new supervisory ethos of power (ibid.: 95). Crucially, I want to argue that while systems of government had been evident in other socioeconomic fields long before, it was not until the 1840s that the atmosphere started to be conceived of, and acted upon, as an object of government. The coming together at this time of Parliament and science in the tentative coordination of air power and knowledge production thus marks an important juncture within which it is crucial to position and interpret our contemporary systems of atmospheric knowledge. No longer was the battle for cleaner air connected to the whim of the Sovereign, or the bourgeois interests of urban elites, it became part of a broader system of government that took as its target the British population, and as its motive the crucial role of the air within the operation of a modern political economy.

If contemporary understandings of, and behaviours towards, British air pollution have been structured by an enmeshing of government and science, it is crucial to consider available conceptual methodologies that can guide interpretations of this complex historical relationship. It is in this context that this chapter leaves the historical narrative that has been briefly commenced here, and will be rejoined in Chapter Three, in order to outline the interpretive and methodological frameworks within which this book is set. The process begins by exploring Michel Foucault's celebrated work on the history of governmental power and reason. While famously distilled in his 1978 lecture series at the Collège de France, *Security, Territory, Population,*

Foucault's interest in governmental forms of power can also be discerned within interrelated lectures and writings (see Foucault, 1998 [1976], 2004 [1997], 2007 [2004], 2008 [2004]). Foucault's analysis of the history of governmental reason and practice is important for this volume on two counts. First it provides one of the most comprehensive accounts of the nature, form and modes of operation associated with liberal forms of governmental power. Foucault's governmental oeuvre thus indicates how it might be possible to discern the role of governmental power and reason within modern atmospheric study, while positioning atmospheric government in the context of the broader orchestration of power and knowledge associated with the modern State. Second, and beyond its role as an ontological test-bed, Foucault's analysis of the history of government also reveals a methodology of power study upon which this book's more specific analysis of atmospheric government is, in part, based. The second section of this chapter introduces and explores key writings in the history of science and science and technology studies (collectively referred to as the sociology of scientific knowledge, or SSK). This chapter does not engage with science studies simply because a history of atmospheric government appears to require an equivalent account of the sciences of the aerosphere. After all, in both his account of the history of government, and early archaeologies of scientific reason, Foucault himself shows a keen awareness of the importance of placing science within the context of both history and power. Science studies do, however, stress the importance of attentiveness to the role of scientific practices and technological things within the development of science, which while implied in Foucault's later historical methods, are often strangely absent from his own analyses.

Atmospheric Governmentalities and Scientific Power: Tracing the Knowledge Effects of *Government with Science*

Science, government and Foucault's apology

On Wednesday 8 February 1978 a flu-ridden Michel Foucault arrived at the Collège de France to deliver the fifth lecture in his courses on themes of *Security, Territory, Population* (see Foucault, 2007 [2004]). Having first apologised for being 'more muddled than usual' (on condition of his flu), Foucault spent the opening moments of his lecture trying to correct an apparent error that was evident in his previous week's lecture. In his famous 'governmentality' lecture of 1 February Foucault described a transition that was evident in eighteenth-century Europe from sovereign systems of power to governmental regimes as 'the transition from an art of government to political science' (ibid.: 106). Yet on 8 February Foucault describes his use of the term science as '[a] thoroughly bad and disastrous word' (ibid.: 116). Although Foucault rejected the notion of a science of government – preferring instead the notion of *political competence* in the acts of governing – the idea that modern systems of government are characterised by a scientific ethos of knowledge production and rational decision making remains a popular mode of characterisation (ibid.: 116).[16] In invoking the idea of a science of government, however, Foucault was not intimating the formation of a scientific epistemology within the State, but describing the spread of a style of economic government that had been popularised by the French school of Physiocrats (such as François Quesnay) in the seventeenth and eighteenth centuries.[17] According to Foucault, the Physiocratic position supported the establishment of a science of government to the extent that it sought the careful calibration of knowledge concerning land, wealth and population (see Charbit, 2002: 860). Yet, the scientific credentials of the Physiocrats were undermined by their unscientific doctrines that suggested the inherent value of agricultural production over industry and rural life over metropolitan living (hence Foucault's apology) (ibid.: 855–6). According to Foucault, such unsupported (and largely unsupportable) views exposed an unscientific, ideological core to the Physiocrats' supposedly scientific programme.

The question that I want to consider in this chapter is whether it is possible to reframe Foucault's invocation of a *science of government* in such a way that it enables us to consider the role of scientific practices and rationalities within the histories of government that Foucault charts. In many ways Foucault's apology for his reference to science seems more to reflect the misgivings that he had about the scientific status of the Physiocratic ideologies of government than an abandonment of science as a context for government formation and reformulation.[18] So what if modern forms of government are not only enthused with an ethos of scientific legitimacy, but are, in part at least, predicted upon scientific methods of knowledge production and analysis? If science is understood less as a series of specialist disciplines (chemistry, biology, physics) and more as a *theory of rules of method* (or methodology),[19] it becomes possible to conceive not so much of a *science of government*, but of a *government with science*. The notion of *government with science* has certain advantages over the idea of a science of government. First, it makes a clear distinction between sciences that claim to understand and justify certain forms of governmental action (i.e. political sciences), and the use of scientific methods within the practices of the State. Secondly, and perhaps most importantly, *government with science* neatly avoids the implication that government can in anyway act as a science. There are many reasons why government can never achieve the status of science. The broad and integrative nature of government means that it can never achieve (or indeed seeks) the kinds of disciplinary expertise associated with sciences (even with the formation of specialist institutional bureaucratic divisions of labour). At the same time the nature of the things that have to be governed within modern society are not always amenable to direct scientific modes of analysis. Thirdly, the political nature of liberal government means that while governments may justify their actions through discourses of scientific knowledge production, such claims tend to be undermined by the ideologies that ultimately inform the governmental programmes that emerge from such knowledge systems. While enabling us to move beyond the idea of a science of government, the notion *of government with science* does emphasise the complex skein that connects the processes of government and the actions of science. Much has already been written on the politics of science and the relationship between State institutions and scientific research (see for example Latour, 1993: 15–29, 2007b; Shapin & Schaffer, 1985). It is, however, important to state that the notion of *government with science* is used here to refer to something more specific than the general mixing of politics and scientific objectivity, the State sponsorship of science, and the governmental organization of experiments. It is about the construction of a scientific apparatus of and for government. This volume is dedicated to exploring the knowledge effects of this scientific apparatus of government; knowledge effects that have been formed between the methodologies of science and reasons for government.

While much of this volume will be dedicated to exploring the popularisation of scientific methods as the basis for governmental knowledge production, it is also important to recognise the epistemological processes that have given rise to a government with science. While not discussed in relation to his reflections on the science of government, in lecture 9 of his 1978 lecture series Foucault reveals the historical context within which the birth of governmental power is connected to the rise of modern scientific techniques and epistemologies (Foucault, 2007 [2004]: 227–53). Foucault charts the emergence of governmental forms of reason in the early modern period in relation to a broad set of processes. The first compilation of processes Foucault discusses concerns the religious upheavals associated with the Reformation. According to Foucault, the anti-Catholic *counter-conducts* associated with the Reformation not only undermined the power of the Holy See, but also brought into question the religious epistemologies associated with the sovereign State (ibid.: 227–236). The Reformation confirmed that there could be alternative theories of how (divine) knowledge could be discerned: a realization that threatened Church and State alike. The second, and most important in the context of this chapter, collection of processes that Foucault connects with the rise of governmental reason are the huge upheavals in epistemology associated with the scientific revolution. According to Foucault, the key impact of the scientific revolution on State power was the formation of a 'great duality' between the sovereign and nature (or the *de-govermentalisation of the cosmos*) (ibid.: 236, 239). While nature had historically been invoked as the justification for sovereign power (i.e. the *natural order of things*), the rise of classical science saw the control over the discourses of nature move from the Leviathan and into the laboratory.[20] Not only did the rise of classical science undermine the epistemological claim of the sovereign to various modes of *justification though nature*, it also gave rise to what Latour has described as the 'multiplication of private spaces where the transcendental origin of facts is proclaimed' (1993: 22). It is the multiplication of these new epistemic spaces of science (including the laboratory, museum and university) that has provided the context within which many have charted the strong divisions that have emerged between the representational realms of politics (society) and science (nature).

Within Foucault's history of governmental reason the rise of science is not only used to describe the demarcation of separate loci of power around the State and science. According to Foucault, the Reformation and scientific revolution resulted in a political system in Europe that was shorn of both God and nature as sources of legitimacy. But it is precisely within the context of this crisis of political rationality that Foucault discerns the emergence of a new form of governmental power, based less on transcendental ideology and more on the *reign of reason* (Foucault, 2007 [2004]: 237). While more will be said on this new governmentality in the section that follows, at this point it is important to emphasise that the new reasons for government that Foucault identifies in seventeenth- and eighteenth-century Europe were based upon systems of collective care and public responsibility. Foucault describes this changing sense of governmental rationality in the following terms,

> [t]he emergence of the specificity of the level and form of government is expressed by the new problematization of what was called the *res publica*, the public domain or state (*la chose publique*). The sovereign is required to do more than purely or simply exercise his sovereignty, and in doing more than exercise his sovereignty he is called upon for something other than God's action in relation to nature, the pastor's in relation to his flock, the father's in relation to his children, the shepherd's in relation to his sheep [...] This is government (ibid.: 236–7).

It is in relationship to the legitimacy of government as a modality of generalised fatherly care, or pastoralism, that States would start to construct a new knowledge-gathering apparatus that could begin to rationally comprehend the entirety of existence they were now to take responsibility for. This new regime of rational care for its people increasingly meant that 'men of politics' conceived of science not so much as an illegitimate threat to authentic political power, but as the technological basis for a new epistemology and methodology of government. It appears that the post-organic, atomised view of the cosmos – developed as part of the scientific revolution – suggested a new way of itemising, calculating and controlling political spaces of various kinds (see Merchant, 1983: 206–15).

Despite Foucault's apology, it is important to note that he explicitly recognises this historical skein of government and science. In his lecture of 8 March 1978, and quoting Chemnitz, he states,

> Certainly *raison d'Etat* has always existed, if by which we understand the mechanism by which states can function, but an absolutely new intellectual instrument was needed to detect and analyze it, just as we had to wait for the appearance of certain instruments and telescopes so we could see stars that existed but had never been seen. 'With their telescopes,' says Chemnitz, 'modern mathematicians have discovered new stars in the firmament and spots on the sun. With their telescopes, the new *politiques* have discovered what the ancients did not know, or which they carefully hid from us' (Foucault, 2007 [2004]: 241).[21]

Chemnitz' wonderfully evocative notion of the *telescopes of the politiques* echoes the idea of government with science I have been attempting to outline in this section. The image of the *telescopes of the politiques* is a metaphor for the ways in which new systems of governmentality were predicated upon the very scientific ideas and methods that initially undermined the legitimacy of sovereign State systems.

It is, perhaps, most appropriate to conceive of government with science as an illegitimate offspring of a State system that is no longer in control of the epistemologies of nature. But it is important to reflect upon two further conceptual aspects of the notion of government with science before we move on. In invoking the idea of a government with science it is important to avoid the tendency to assume that there is an homogeneous entity termed 'science' which can be neatly colonised by State authorities. At a practical level, Hacking reminds us that what we routinely refer to as science is actually a collection of disparate practices that include abstract mathematical induction, puzzle solving, experimental discoveries and clarifications, technological design and advancement, and the development of systems of refined measurement (2005 [1983]: 7–9). Beyond the practices of science, there has been a long historical debate concerning what constitutes legitimate scientific methodologies (ibid.; Popper, 2002 [1950]).[22] In the context of the various practices and methods deployed by scientists, it is crucial to establish the key features of the science we might expect to be implicated within a *government with science*. While, at one level, this volume does consider the role of State institutions within the support for specific forms of atmospheric experiment, and acknowledges the role of scientific discovery in reshaping governmental rationality towards air politics, government with science is best conceived of in relation to the seemingly mundane deployment of scientific techniques and technologies within the refined measurement of objects of governmental concern. In using the term techniques of science, however, I am not merely referring to the actual tools and procedures involved in the collection of empirical data. Government with science is about a style of government that deploys the tools of science, but also applies scientific systems of knowledge classification, comparative ordering and compilation. While routinely overlooked, in this volume I argue that techniques and technologies of measurement have been pivotal to the emergence and changing historical dynamics of governmental power. They have, to use more dramatic terminology, offered the practical basis for the historical shift from the transcendental State of God and nature, to the empirical State of government.[23]

The second and related clarification of the term *government with science* I wish to make pertains to its use as an historical marker. There is an implicit danger in deploying terms such as government with science that they can be seen and used as devices of historical demarcation around which beliefs in pre-scientific government can be formed and consolidated. But the term government with science is neither meant to suggest that forms of government have ever been completely isolated from scientific reason, or that in the modern world the rationalities of government are driven by scientific methods alone. It is clear that pseudo-scientific principles, from the pre-enlightenment period, informed government ideologies and practices for a considerable length of time. It is also apparent that many modern states have adopted *anti-scientific*[24] postures to support their ideological stances, or manipulated scientific procedures as a basis for constructing the historical absolutisms upon which authoritarianism thrives.[25] It is against this historical record that government with science is not deployed as a mechanism of temporal demarcation, but as an indicator of historical tendency: an historical tendency for objects of government to be increasingly identified and constructed through the evolving methods of empirical science.

Histories of government and Foucault's 'little experiment'

In this section I want to move beyond Foucault's discussions of science to focus more explicitly on his motives and methods when studying governmental history. Many critical reviews of Foucault's analyses of governmental power and history already exist (see for example Burchell, Gordon & Miller, 1991; Dean, 1999; Elden, 2007; Hannah, 2000; Legg, 2007: 1–28; Rose, 1999a: 15–60). In this context, I do not wish to repeat the now relatively familiar genealogy (or more often rigid archaeology) of Foucault's governmental ideas. Instead I propose to do two things: (i) to explore

Foucault's histories of government in specific relation to their relevance to the history of atmospheric power and knowledge; and (ii) to consider a latent paradox in Foucault's governmental histories pertaining to his commitment to the study of the *practices* of government, but his tendency to excavate governmental *reason*. I argue that this relationship between the study of governmental practices and reason has important implications for how we begin to conceive of the study of government with science in the shadow of Foucault's oeuvre.

The fulcrum of many accounts of Foucault's analysis of governmental power is his concept of *governmentality*. While time is spent in this section explaining and unpacking the notion of governmentality, I want to argue that if taken in isolation this concept does not reveal the full intention of Foucault history. In lecture 4 of his 1978 lecture series Foucault does assert his desire to construct a 'history of "governmentality"', yet at the very end of his course Foucault offers an interesting reinterpretation of his own reasoning,

> All I wanted to do this year was a little experiment of method in order to show how starting from the relatively local and microscopic analysis of those typical forms of power of the pastorate it is possible, without paradox or contradiction, to return to the general problems of the state, on the condition that we [do not make] the state [into] a transcendent reality whose history could be undertaken on the basis of itself (Foucault, 2007 [2004]: 358).[26]

I want to argue that if taken as 'a little experiment of method', Foucault's governmental histories are not only suggestive of an ontological condition of power (governmentality), within which atmospheric histories could be set, but also of a methodological framework through which the study of atmospheric government could usefully be conducted.

At the commencement of his 1979 lecture series (*The Birth of Biopolitics*), Foucault clarifies his particular historical approach to government,

> [c]hoosing to talk about or to start from governmental practice is obviously and explicitly a way of not taking as *a priori*, original, and already given object, notions such as the sovereign, sovereignty, the people, subjects, the state, civil society, that is to say, all those universals employed by sociological analysis, historical analysis, and political philosophy in order to account for real governmental practice [...] In other words, instead of deducing concrete phenomena from universals [...] I would like to start with these concrete practices and, as it were, pass these universals through the grid of these practices (Foucault (2008) [2004]: 2–3).

Foucault's method is thus focused on the microscopic aspects and practical manifestations of governmental power. He utilises these methodological tools to insulate his analysis of government from the potentially overpowering explanatory dynamics of existing sociological categories. Let us now consider the ontological and methodological implications of Foucault's governmental histories in turn.

Governmentality and histories of power

Described by Foucault as his 'ugly word', the term governmentality does not actually occur until the fourth lecture of his 1978 *Security, Territory, Population* course (Foucault 2007 [2004]: 108). It is consequently important to interpret governmentality in relation to Foucault's earlier lectures of his 1978 course. In locating governmentality within these preliminary lectures, however, it also becomes possible to situate the idea in the context of Foucault's broader engagement with questions of power. At the commencement of the 1978 course Foucault clearly positions his account of governmental history in relation to his ongoing interest in something he terms as *biopower*.[27] According to Rabinow and Rose, biopower is an expression of power that takes human life as its target (2006). As an *administration of life*, the notion of biopower is a familiar expression of governmental practice in the world today. While the power of the State continues to find partial expression through the legitimate exercise of violence and repression, we also experience government in everyday life as a set of institutions (including hospitals, clinics, schools and dietary advice bureaus) that seek to support and enhance our biological fecundity.[28] In two lectures presented at the State University of Rio de Janeiro in October 1974 Foucault began to outline an account of biopower in relation to an historical analysis of the history of State medicine and public health.[29] In the second of his Rio lectures, entitled *The Birth of Social Medicine*, Foucault traces a transformation in what he terms *biohistory* (or the impact of medical practice on human history) (Foucault, 2000a [1994]: 134). According to Foucault, from the eighteenth century onwards it is possible to discern the transformation of medicine from something that is the narrow concern of the medical expert, to an arena of public concern and strategic State intervention. Suddenly the health of the body was not merely a private concern, but something that was open to regulation through the governmental control of public environments and the territorial orchestration of medical facilities.[30] While it is tempting to ascribe biopower to a kind of political medicine that is institutionally removed from the human body and resides in abstract statistics of health and strategies of urban economic development, Rabinow and Rose remind us that biopower exists at the intersection of the anatomical body and the statistically delimited population (2006: 196). As a strategy of power that seeks to *govern each and all* within the administration of life, biopower is essentially a context within which it is possible to discern the rise of a whole range of 'great technologies of power' (including sexuality, economics and even nationalisms) (ibid.).

The notion of biopower provides an important conceptual context within which this volume's analysis of atmospheric government is situated. As the most immediate requirement for human life, the quality and availability of breathable air has clearly been a crucial environmental medium in and through which biopower has been expressed. In his discussion of the birth of social medicine Foucault identifies the regulation of the air as one of the central priorities of the new biopolitical regime of power (Foucault, 2000a [1994]: 147–8). He consequently describes how, during the eighteenth and nineteenth centuries, governmental authorities took an increasing interest in the role of the air in determining illness, '[b]ecause of its excess chillness, hotness, dryness, or wetness' and its role in carrying disease (ibid.: 148). While not talking explicitly about air pollution, Foucault charts how, in the name of urban medicine, eighteenth- and nineteenth-century urban planners, architects and medical officials sought to *ventilate* cities through the formation of *aeration corridors* and atmospheric currents (ibid.: 148).

Returning to the 1978 lecture series, Foucault argued that governmentality and biopower were connected to the extent that in order to understand the emergence of *power over life* it was necessary to first chart the historical rise of governmental power. The connection between government and biopower can be discerned in two key ways. First, as mentioned previously, the notion of government is synonymous with an ethos of care, an ethos of care that biopower's concern with the fostering of life, rather than the determination of death, clearly echoes. Second, the expanded institutional scales of power associated with State-based systems of government (whether at the level of a city or nation) appear crucial to the aggregate management of life through populations that Foucault associates with biopower. To these ends, Foucault developed his account of the history of government in order to explain a form of contemporary (bio)power he had already identified. But if Foucault sees his history of government as a vital condition for interpreting biopower, much of the early lectures of Foucault's 1978 lecture series are dedicated to the exploration of modalities of power that help to explain and conceptualise the rise of governmentality. According to Foucault, historical power has been characterised by three primary modalities: sovereignty, discipline and security (Foucault, 2007 [2004]: 5–6). Through discussions of varied examples – including the regulation of the plague, urban planning and grain shortages – Foucault explains the key historical characteristics of these different mechanisms. Accordingly, he describes sovereign modalities of power as those typically associated with the medieval period, and involving strong legal prohibitions and pronouncements on what can be done where and when (ibid.: 9–10). In contrast, Foucault argues that disciplinary mechanisms of power, which came to prominence during the eighteenth century, are less about prohibition and more concerned with the careful surveillance and supervision of socioeconomic life (ibid.: 10). Finally, Foucault claims that the contemporary era is characterised by a new mechanism of power, the power of security. Foucault claims that the society of security is based upon a less overt form of power that attempts to supervise tolerable *bandwidths of existence*, only intervening when optimal socioeconomic conditions are under threat (ibid.: 6).

Although Foucault emphasised that the mechanisms of power he outlines in the early lectures of 1978 should not be interpreted as discrete geohistorical power regimes, with sovereignty being replaced by discipline, which in turn is replaced by security,[31] he does argue that there is an historical tendency towards mechanisms of security within histories of power (Foucault, 2007 [2004]: 107–8). In his discussion of different modalities of power Foucault ultimately entreats us to interpret '[a] triangle: sovereignty, discipline, and governmental management, which has population as its main target and apparatuses of security as its essential mechanism' (ibid.: 107–8). It is in this context that although this volume focuses predominately on the actions and mechanisms of governmental security on the construction and management of British air pollution, analysis remains sensitive to the continued deployment of sovereign and disciplinary techniques in British air politics and to the impacts of these mechanisms on the nature of atmospheric governmentality. To put it another way, in the chapters that follow it will be possible to see that tactics of sovereign control over air pollution did not stop with the proclamations of King Edward in the fourteenth century and the purported execution that followed (see previous chapter). Instead, it will be argued that the tactics of sovereignty have become part of a broader shift towards systems of governmental security within which atmospheric control becomes more closely tied to a power over collective life as opposed to a power over death.

Having established that governmentality as a concept is connected to the allied notions of biopower and mechanisms of security, it now important to reflect in greater detail on precisely what governmentality is. Foucault himself provides us with a tripartite depiction of the primary features of governmentality. First, Foucault claims that governmentality is characterised by an *ensemble of techniques of calculation and assessment* dedicated to the regulation of the population and with the techniques of security as its primary mode of action (ibid.: 108). Second, governmentality is taken to refer to an *historical tendency* in and through which governmental power (and associated institutions and knowledge production apparatus) have come to prominence (ibid.: 108). Third, governmentality is taken by Foucault to denote a *process of State transformation* whereby States that were previously dominated by techniques of sovereignty and disciplinary power are governmentalised (ibid.: 108). When articulated as simultaneously an *ensemble of techniques*, an *historical tendency*, and a *process of State transformation* it could be argued that Foucault's attempt to unpack governmentality serves only to further obfuscate the elusive concept. Margo Huxley encourages us to understand governmentality as having two basic elements: (i) the practices, policies and programmes of 'government'; and (ii) the 'mentalities' or 'thoughts' that provide the rational basis for the practices of government (Huxley, 2006). In a further search for clarification, Matthew Hannah entreats us to conceptualise the elusive concept of governmentality as a form of nationalised biopower (Hannah, 2000: 23). But care must be taken in this context not to equate governmentality too closely with an upscaled version of biopower.[32] It is for this reason that I interpret governmentality as the rise of a national apparatus of security, or, to put it a different way, the emergence of a system of government that is able to function at the level of a national social economy.

The relationships between governmentality and biopower are historically and conceptually complex. As both biopower and governmentality appear to reflect large geohistorical shifts in the nature and intent of power, it would be unwise to suggest that the rise of governmental forms of power caused the emergence of biopower, or indeed vice versa. What it is possible to argue is that the rise of spatially extended systems of governmental administration and knowledge production during the eighteenth and nineteenth centuries increasingly enabled biopower to be expressed at the level of the population as whole – rather than on the basis of the individual body, or urban community. It was in recognition of this fact that while still ultimately exploring the nature of biopower, in 1978 Foucault sought to chart a history of government, as opposed to developing a dedicated history of biopolitics (a task he undertook in his lecture course at the Collège de France the following year (Foucault, 2008 [2004])). This is also precisely why after outlining his neologism that is governmentality, and having situated it in relation to questions of biopower and mechanisms of security, that Foucault dedicates the majority of his course on *Security, Territory, Population* to a history of government. Reflecting briefly on Foucault's history of government is important on two counts: (i) because it provides an insight into the particular modes of political rationality that characterise a desire to govern the atmosphere and which occupy the primary focus of this volume; (ii) because it reveals a missing link in Foucault's own historical method which has particular resonance for how we might begin to understand the role of science in governmental history. Let us now consider Foucault's account of the history of government in greater detail.

Government in historical text and context

We have already discussed the religious and scientific upheavals that Foucault associates with the rise of governmental forms of power in the seventeenth and eighteenth centuries in the previous section, but Foucault argues that it is possible to trace the rationalities and practices of government to much older points of historical origin. Although the verb *to govern* has a range of contemporary and historical meanings,[33] if we interpret the acts of governing as procedures dedicated to establishing a certain order in the arrangement of things that are necessary to modes of socioeconomic survival, Foucault argues that it is possible to trace the ideas and practices of government to the pre-Christian East and the formation of a pastoral form of power (Foucault, 2007 [2004]: 123). By locating the origins of governmental thought and practice within early pastoral forms of power, Foucault equates the processes of governing not with the relationship between a monarch and his subjects, but between a shepherd and his flock. There are two important aspects of pastoral forms of power that connect it with contemporary forms of governmental action. First, it is associated with the exercise of power over a 'multiplicity on the move' (ibid.: 126). As with the shepherd overseeing his flock as it feeds and roams from pasture to pasture, government does not involve tight spatial control on movement and exclusions from space, but a desire to enhance circulations of people, goods and the various forces of life. Second, Foucault discerns in pastoralism a distinctively beneficent manifestation of power (ibid.: 126). Although, as Hannah (2008) has recently argued, the forms of biopower associated with modern government are not always motivated or directed by a sense of *biophilia* (love of life), it is clear that government draws much of its legitimacy from a sense of the collective care its affords its population (i.e. the State as *good shepherd*).

Having located the genesis of governmental thought and practices in the pastoral regimes of power associated with the pre-Christian East, Foucault devotes significant time to charting the rise of pastoral techniques of power within Christianity and subsequent secular philosophies and political treaties. Ultimately Foucault describes how the governmental mode of pastoral power was gradually adopted as a new rationality for State action and legitimacy during the seventeenth and eighteenth centuries.[34] In asserting the pre-eminence of governmental techniques of power Foucault is able to interpret modern forms of State power as merely a moment within the longer history of government: a period within which government is practised at the level of a territorial population. But, despite the undoubted importance of the history that Foucault outlines, there is a tendency within his history of government to focus primarily upon the discourses of governmental reason as they appear in key philosophical texts, religious teachings and political treaties. While this analysis provides great insights into governmental reason (mentalities), it provides only a limited account of the local and microscopic practices of government that Foucault himself asserts are so central to his historical method of analysis. To put things another way, we only see half of the analytical promise of governmentality identified by Huxley (2007). While Foucault's concern with historically specific and politically contingent rationalities of government continues to insulate him from *a priori* forms of historical explanation, it does not offer an insight into the material conditions under which the reasons for government change.

There are two reasons why Foucault's tendency to focus on the historical rationalities, as opposed to practices, of government are significant for the analysis presented in this volume. First, in exploring the particular histories of atmospheric government this volume does not wish to see the rationalities of government as a fixed context within which the practices of air politics and science are carried out. Instead analysis attempts to reveal how certain rationalities of air government were actively shaped and transformed at the practical interface of atmospheric science and air government. It is not just that the study of practice provides a touchstone of reality against which to test the historical presence of abstract rationality. Instead Foucault intimates that the role of rationalities in shaping historical conduct and action can only ever be discerned by the historian at the level at which they are tried, tested, adapted and ultimately accepted or rejected (Foucault, 1991: 81–2). The second, and related issue, raised by the relative absence of government practices within Foucault's analysis is the impact this approach could have on the effective construction of a history of government with science. While the notion of science carries with it its own sense of rationality, the presence of a scientific method as a grounds for governmental knowledge production is not defined by a set of fixed techniques of truth telling, but on the basis of an open commitment to ever-changing forms of practice, technique and technology that can, at any point, falsify governmental truth and undermine mentalities of the State. A history of government with science thus requires an historical methodology that is able to study the co-evolution of governmental rationality and practice. This is a Foucauldian methodology, but one only sparingly deployed within his account of the history of governmentality.

Governing the subject and cultivation of self-governance

Before returning to discussions of the interconnected histories of science and government, I want to briefly reflect upon an additional aspect of Foucault's analysis of governmentality that has significant bearing on the analysis of atmospheric government presented throughout this volume. While much of Foucault's analysis of governmentality charts the rise of government systems that exercise power at the level of aggregate phenomena like populations, national territories and atmospheres, one of the most original aspects of his work (particularly in comparison to other streams of State theory) is his concern with relationships between governmental power and the individual subject of power. According to Foucault, the various acts associated with governmental power can be summarised in one simple formulation: government is the *conduct of conduct* (see Foucault, 2007 [2004]: 121).[35] Put simply, this formulation reminds us that in order to establish the *right disposition of things* – in relation to a functioning social economy at least – the processes of government do not simply act on things. To govern, at least at aggregate levels, requires the effective cultivation of subjects who are able to act on things in ways that are commensurate with governmental goals (what Foucault terms *obedience as a unitary*) (ibid.: 174). It was in this context that Foucault was keenly aware of the significance of the subject as a site of governmental strategy. At one level the cultivation of subjects who are capable of certain styles of conduct reflects the ultimate limitations of both sovereign absolutism and disciplinary surveillance. Neither the all-powerful sovereign, nor the society of the Panopticon, can account for the actions of all individual subjects at all times. At another level, however, the construction of subjects capable of government reflects the paternal desire of good government to oversee the activities of its population in the same way that a father cares for his household (ibid.: 105). In this context the coordination of subjective action, which is suggested by the notion of the *conduct of conduct*, brings into question how the project of government becomes an internalised norm within the everyday action of citizens.

According to Foucault, the key to understand how and why modern government is able to extend its project into the realms of personal conduct and decision making is an appreciation of the historical origins of governmental forms of power into the Christian pastorate (ibid.: 163–90). As Foucault notes, and despite the emergence of secular societies, the rise of the Catholic Church marked the birth of a form of governmental power from which we are still to escape (ibid.). This is a form of power and moral influence that is able to reach into the most intimate and immediate everyday decisions and practices of individuals and impose its dictates through the very soul of the subject (in 'every moment of their existence' (ibid.: 165)). According to Foucault, it was the Christian Church, with its combination of heavily institutional power and the pastoral ability to reach into the deepest spheres of personal conduct, which provided the paradigm of power for modern government. Although modern systems of government reflect the forms of caring control first seen in the Christian pastorate, it is important to recognise one key difference between forms of liberal governmentality and pastoral power. While the Catholic Church sought to reach into the intricacies of personal conduct in order to establish a form of pre-ordained moral order in the subject, Foucault charts how modern government works with the pre-existing desires, wants and needs of the subject and seeks to cultivate these in the pursuit of broader mosaics of socioeconomic order. To these ends modes of governmentality are perhaps best thought of as systems for *acting on existing actions*, rather than disciplining errant behaviour. Crucially, however, the cultivation of a governmental self is predicated on the ability of individuals to recognise the relationship between their own conduct and welfare and the broader well-being of the nation in which they live (see Foucault, 2000b [1979], 2000c [1979]).

Foucault's later work on the care and cultivation of the self recognises that systems of self-government and personal control often emerged in the absence of strong authoritarian systems of power (Foucault, 1990 [1984]: 41). The more recent work of Rose has developed on this Foucauldian theme in a more contemporary context (Rose, 1999b). According to Rose, the emergence of psychology and psychiatry in the nineteenth and twentieth centuries marked the birth of a new set of sciences that were devoted to understanding the actions and motivations of the human subject. Rose's work highlights how the production of a governable subject is not a product of the formal, hierarchical institutions of a government alone, but is ultimately realised on the basis of subjects recognising themselves as sites for contemplation, reflection and self-regulation. Throughout this volume attention will be placed on the connection between emerging systems of air government and the fostering of new forms of atmospheric self. Significantly, accounts of the emergence of new forms of atmospheric self-conduct will remain firmly connected to the terrains associated with the study of government with science which define this volume. Through a study of various theatres of atmospheric persuasion (including exhibition halls, lecture theatres and web sites) analysis reveals that the formulation of new systems of atmospheric conduct in Britain were in part based upon the promotion of new modes of 'scientific' practice within the home and workplace.

Histories of Scientific Knowledge, Practice and Technology

It was, after all, Einstein who famously said that we should take little heed of scientists' formal reflections on what they do; we should instead 'fix [our] attention on their deeds'

Shapin, 2001: 106

Preliminary perambulations on governmentality and the social study of scientific knowledge

In this section I explore the valuable insights that a body of work collectively referred to as the sociology of scientific knowledge (hereafter SSK) can offer to the general study of government with science, and the more specific analysis of the government of British air pollution. As we will see, while this body of work has clear methodological associations with Foucault's own genealogies of government, it provides important additional perspectives on the nature of scientific and political change in history. It is important to state initially that SSK does not constitute a clearly defined or even internally consistent disciplinary tradition. SSK emerged during the 1970s as a predominantly British intellectual project and brought a new set of scientific objects of enquiry into the sociological tradition, while introducing a novel set of sociological methods to the study of scientific practice and knowledge production (Shapin, 1995: 289). Following Steven Shapin, I take the idea of SSK to be broad enough to include both the fields of science and technology studies (STS) and work within the history of science.[36] According to Shapin the primary concern of SSK is, '[h]ow to interpret the relationship between the local setting in which scientific knowledge is produced and the unique efficiency with which such knowledge appears to travel' (ibid.: 290). The association between SSK and the locating of science within specific sets of corporeal practices, laboratory settings and institutional structures is perhaps the most popular context within which people think about a sociology of science. The desire to position the production of scientific knowledge within different, but highly specific, locales is perhaps most famously associated with Haraway's call for a more situated sense of knowledge (Haraway, 1991).[37] While I want to argue that scientists have a more situated sense of their self and their practices than Haraway suggests, I also want to claim that situating and locating science is a vital process in attempting to understand the entwining of science and government. The second aspect of Shapin's account of the key objectives of SSK – namely a concern with the ability of scientific knowledge to travel – is perhaps associated far less with this academic subdiscipline. Yet a concern not only with the place of scientific knowledge production, but also with its apparent mobility, raises crucial questions about the transformation of scientific knowledge from the particular to the general; the specific to the universal. Throughout this volume I argue that the relative ability of knowledge to travel has a crucial role in determining what is classified as scientific knowledge and what is disqualified as inauthentic and vernacular. Furthermore, I claim that the particular qualities that enable scientific knowledge to attain a relatively high level of mobility are precisely why such knowledge is so important to the practices and strategies of government. The following section reflects in greater detail on the conceptual concerns and methodological parameters of SSK. The remainder of this section, however, considers how and why it is helpful to consider Foucault's analysis of governmentality alongside work within SSK.

There is always a danger when two (or more) bodies of work appear to address the same set of problems of assuming two things: (i) that their common focus of analysis can outweigh any epistemological or methodological difference when assessing their compatibility; and (ii) that significant extra analytical value can be gained by combining the insights of the different intellectual approaches. The fact that Foucault asserts the importance of scientific rationality within his history of government without providing a detailed account of the role of scientific practices within the State, while SSK provides a rich body of historical analysis of the practical relationships between science and politics, could easily lead us to unthinkingly, but conveniently, adopt these two assumptions. Actively engaging with and questioning these common assumptions is, however, complicated in this instance by the fact that there is no single, or representative, epistemology within SSK.

Placing epistemological issues to one side for the moment, perhaps the clearest common ground that SSK and Foucault's histories of government share is their commitment to the study of practical histories. As Shapin argues, SSK is dedicated to the study of what scientists do, as opposed to the analysis of the meta-scientific discourses which science utilises to explain its methods. In a similar sense, Foucault is keen to move beyond the ideologies that undergird the cold monster of State-based explanation to provide a micro-history of governmental techniques and struggle. Perhaps the clearest indication of the methodological commonalities of Foucault's genealogical methods and SSK is expressed by Foucault when he describes his genealogical method as being *anti-scientific*, or at least against certain trends within the scientific structuring of knowledge (Foucault, 2004 [1997]: 9).[38] As with SSK then, it is not so much that Foucault's historical methods are against science per se, but rather that they seek to expose the complexities of knowledge production that can be hidden beyond the veneers of scientific logic and absolutism. It is in this context that Foucault asserts the importance of studying subjugated knowledge that is 'disqualified by the hierarchy of erudition and the sciences' (Foucault, 2004 [1997]: 8). What methodologically unites Foucault's genealogies and the work conducted in the name of SSK is a realisation that to study practical histories is to provide analytical perspective on the strategic, but often arbitrary, categorisation of some knowledge as science and others as sub-scientific. To study subjugated knowledge thus brings into focus a sense of the historical struggles that can be lost within the narratives of meta-science: struggles that involve the mixing of knowledge and power, reason and force, freedom and suppression.

Having established the common methodological ground of governmentality and SSK, it is important to consider the second question that I posed: even if two approaches have much in common why is it necessary to even try to combine them? While notions of governmentality and SSK suggest the importance of focusing on the subjugated knowledge of scientific and political history, SSK as an intellectual movement has developed a particular set of techniques for discerning and interpreting the complexities associated with the production of scientific data. Within Foucault's oeuvre it is possible to discern a desire to subvert the hierarchies of learning associated with formal science by deliberately studying what he termed pseudo-sciences (such as psychology, demography and even government). Through the study of pseudo-science – or sciences still in the process of attaining scientific status – Foucault was able to expose the practices and processes that had to be suppressed in order for knowledge to attain scientific standing. Unlike the work of Foucault, SSK has emerged as a subdiscipline that is specifically devoted to analysing the operation of politics within all sciences. Consequently, in addition to attempting to understand and explore the leftover knowledge of the scientific method, SSK requires an ability to study the internal workings of sciences as a precondition for understanding the strategic demarcation of legitimate and illegitimate knowledge. Thus, while Foucault's genealogies move through the fertile outskirts of scientific knowledge, analysing the production and subjugation of 'non-scientific knowledge' by sciences-in-the-making, SSK requires us to move into the centre of established scientific methods and epistemologies: it requires a commitment to study the intimate technicalities of sciences (see Shapin, 1995). In an interesting reflection on SSK, Steven Shapin claims that one of the problems of gaining and retaining students within this subdiscipline is that they must be competent in both social and scientific analysis, philosophy and empirical science, theory and technological procedure (ibid.: 293). SSK appears to require the presence of certain scientific competences that are not necessarily required within a genealogical approach. Consequently, although SSK is devoted to interpreting science within a range of sociological contexts (including cultural traditions, institutional procedures and political power struggles) it does not propose an analysis of science in practice that commences outside of the actions of scientists (perhaps with discourse): SSK starts with an interpretation of the mechanics of science as a basis for understanding how, when and where the sociological becomes a factor within scientific method. But SSK does not only involve attentiveness to the things of science that go beyond the discourses of scientific method, it also stresses the importance of considering the role of a series of technological devices and procedures, chemical compounds and biophysical processes as agents within political and social history. A commitment to study such a range of actors (or *actants*) within the constitution of scientific history and practice is not, however, an obdurate dedication to a form of naive materialism, rather it suggests an analytical obligation to the full range of conditions under which scientific knowledge is produced, fabricated and contested (see Latour, 1999, 2006). I claim that it is this attentiveness to the empirical details of scientific practice, action and change that means SSK can bring significant methodological insight to analyses of the history of government with science.

Reflections on the State and government in SSK

By way of further introducing the key methodological and epistemological relevance of SSK to the study of governmental history in general, and atmospheric government more specifically, I now want to reflect upon prominent studies within the SSK tradition that speak directly of the relationships between State, government and science. This is an important analytical move because it exposes some of the reasons why an SSK approach to historical analysis may be conceived as incompatible with the theories and methods of governmentality, but also helps to reaffirm the common ground that exists between the study of scientific and governmental history.

Perhaps the most celebrated analysis of the relationship between science and government to emerge from SSK is Shapin and Schaffer's *The Leviathan and the Air-Pump* (1985). *The Leviathan and the Air-Pump* is of significance to the analysis presented in this volume not only because of the insights it provides into the role of SSK within the analysis of governmental history, but also because of the overt attention it draws to the politics of the air. Although Shapin and Schaffer's study is dedicated to exploring what Latour (1993) describes as *politicoscientific mixtures*, by focusing its analysis on emerging scientific and political modes of knowledge production in seventeenth-century England the volume provides a unique insight into how and why a sense of separation was established between the spheres of sciences and the realms of politics and government. In this context, Shapin and Schaffer chart the emergence of the experimental spaces associated with the Royal Society and the work of Robert Boyle in the seventeenth century and how they clashed with the visions of governmental knowledge production associated with Hobbesian political philosophy. While the experiments of Boyle suggested that the study of nature and natural forces (the realm of the experimental scientist) were distinct from the study of human affairs (the realm of the State), Hobbes' political philosophy suggested that the new spaces of experimental scientists were dangerous zones where the influence of special interests could corrupt and destabilise the harmonious certitudes of the sovereign (Shapin & Shaffer, 1985: 333, 337).

While *The Leviathan and the Air-Pump* provides valuable insights in to the historical origins of the concerns that inhabit our consciences when we speak of the State and science, Shapin and Shaffer's detailed study of the scientific and political philosophies of both Boyle and Hobbes leads them to conclude that it is futile to attempt to construct histories of science that are devoid of the political, or of the State that ignore science. Shapin and Schaffer consequently discern the common ground of the history of science and politics in three ways: (i) through the construction of a political sphere and community within which science operates and scientists negotiate their everyday practices; (ii) through the evident and increasing use of knowledge produced by the polity of science within the practices and tactics of States; and (iii) on the basis of the *conditional relationship* in and through which the nature of scientific practice and the nature of governmental technique are intertwined (ibid.: 332). While this volume will, at different times, reflect upon the politics of scientific practice, and the indirect deployment of scientific discoveries within government, it is the *conditional relationship* that Shapin and Schaffer identify between the norms of science and the structure of political society that appears to have most relevance to the historical analysis of government with science that provides the focus of this book. In broad geohistorical terms, *The Leviathan and the Air-Pump* reveals how the emerging political ideologies of liberalism and the social contract found ideological succour from the personal freedoms associated with experimental science, while the political structures associated with liberal government provided a supportive context for the development of modern empirical science (ibid.: 343). In this volume analysis seeks to develop upon Shapin and Schaffer's study of the conditional relationship between the structures of the liberal polity and the knowledge products of experimental science in order to explore how the rise of governmental ideologies of the State initiated a novel set of knowledge production practices at the intersection between State and science. Ultimately analysis will show that the totalities of knowledge required to operationalise governmentalities would involve an intensification of the relationships between science and the State, an increase in the importance placed on scientific knowledge and practice within political decision making, and a heighten governmental orchestration of scientific activity. In essence analysis considers the forms of scientific practice and organisation that were fostered and supported by (and in turn sustained) emerging systems of governmentality.

The second key exploration of the relationship between governments and science to emerge out of the SSK tradition is Latour's (1988 [1984]) *The Pasteurization of France*. Although Latour's analysis of Pasteur's experiments does not focus explicitly on the relations between State and science, in his desire to understand how a set of scientific practices take on national significance Latour provides invaluable insights into the spatial dimensions of government with science. Latour's analysis of the spread of the techniques and methods of pasteurisation is significant because he analyses the geographical dispersal of scientific methods and practices without assuming the *a priori* ability of scientific truth to empower such spatial movement, or the pre-given ability of the State to effortlessly institutionalise science at a national level. Consequently, in his analysis of the nationalisation of scientific technique, Latour simultaneously problematises the nature of both political and scientific power. In relation to science, Latour's study of pasteurisation emphasises that Pasteur's germ theory, and the attendant practices of bacteriology, did not take effect because of the overwhelming will and power wielded by Louis Pasteur himself. In opposing the conventional view of scientific history that position the *great women and men of science* at the epicentre of scientific change, Latour reveals that in order for science to travel from its local point of origin, discovery or demonstration, scientists have to engage a whole range of actors, institutions and things. Latour reflects,

> Why should we still do for Pasteur's genius what we no longer do for Napoleon's or Rothchild's? [...] they almost invariably suppose that where science is concerned, the diffusion of an idea, a gesture, a technique, poses no particular problem; only the constitution of the idea or gesture is problematic (Latour, 1988 [1984]: 15).[39]

As Shapin reminds us, it is the movement of science, as much as its embodied localism that necessitates an account of politics within scientific history (Shapin, 1995: 308). For France to be pasteurised thus required Pasteur, and his many assistants and supporters, to persuade other scientists of the value of his methods. It also required the support of divisions of government to effectively train and mobilise hygienists to act on the disease prevention methods suggested by Pasteur's analyses. To acknowledge the politics of pasteurisation is not to question its scientific validity, or indeed the truth of Pasteur's microbiological manipulations, but to recognise the role of political engagement in shaping what it is possible for scientists to achieve. It is in this context that Latour entreats us to abandon our ideological squeamishness about mixing accounts of science and politics (see Latour, 1988 [1984]: 7).

Perhaps what is most important about Latour's analysis of the pasteurisation of France are the insights that it provides into what actually makes something scientific. Latour's answer to this question is, I think, significant because it provides insight into the inevitable relationships that are continually formed between sciences and governments of different hues. According to Latour, what makes knowledge scientific are the mechanisms of technological and institutional support that re-enforce its movement and validity. To become a scientific practice methods of knowledge production have to constitute technologies (such as the thermometer, microscope or barometer) that are in the words of Latour both *obligatory points of passage* for the production of accepted knowledge, and the *immutable mobiles* in and through which scientific practice can be dispersed in both time and space (ibid.: 43–9). Reflecting on the work of Latour, Shapin astutely concludes that, 'When all the elements in a network act together to protect an item of knowledge, then that knowledge is strong and we come to call it science (Shapin, 1995: 308). Throughout this volume I argue that structures of government have provided a crucial context within which the metrology of scientific standards has been supported. As Andrew Barry observes, governments are constituted not purely on the basis of the demarcation of geographical territories, but also through the formation of *technological spaces* of compatible technique and procedure (Barry, 2001: 3). To these ends, this volume interprets the science of atmospheric pollution as a set of practices and technical procedures produced at the intersection of government and local scientific practice. This is not to say, as we shall see, that science operates on the basis of its insertion into the circulatory capacities of territorial government. Government with science appears to be as much about the scientisation of governmental bureaucracies as the governmentalisation of scientific networks.

Having considered Latour's account of the nationalisation of Pasteur's laboratory I am conscious that a question I raised earlier in this section appears to represent itself: namely why is it necessary to combine Foucault's theories and methods of governmentality with the theory and methods of SSK? After all, within the work of Latour we are presented with a microstudy of science and how it is nationalised. Furthermore, Latour's approach and concepts can provide significant purchase of the study of British atmospheric government with pollution science. Indeed, as we move through this volume it will be possible to discern key networks, obligatory points of passage and immutable mobiles through which such a government with science has emerged. I do not, however, utilise Latour's concepts explicitly within this volume for two main reasons. First it is clear that, notwithstanding the wide deployment of Latourian concepts within contemporary social science, Latour's oeuvre is not a theoretical toolkit in and through which to interpret reality. Rather Latour's approach to the history of science, and the politics of society and nature, is more akin to a kind of methodological posture (see Latour, 2006). This methodological posture is grounded on the abandonment of the discourses of both the natural and social sciences in favour of the radically open *tracing of associations* (Latour, 2006: 5).[40] It is in this context that Latour rejects the ability of theories of the State (and by definition forms of government) to provide *a priori* structures of explanation, or even objects of analysis (Latour, 1993, 2007b; Callon & Latour, 1981). While I am sympathetic to Latour's methodological intent I am, at least at a practical level, compelled to deploy Foucault's notion of governmentality as a methodological and theoretical trajectory for this project. To this end I seek to avoid the potentially infinite regress of a radically open history of science and government suggested by Latour, by asserting that the science of atmospheric pollution in Britain is a science, in part, forged within the forceful historical dynamics of governmental reason and associated apparatus of care that have been an emerging feature of social organisation for millennia. This does not, however, preclude a commitment to the indeterminacy of the relationship between atmospheric government and science as they have co-evolved over the last 200 years. This is a commitment that I believe can be well supported by the kind of dedicated attentiveness shown to the role of the micro-practices and things of science in SSK. The question then remains what type of air pollution sciences have been nurtured by systems of liberal (and neo-liberal) government, and what role have changing forms of atmospheric science had on the structure and reasoning of air government?

Conclusions: On Histories and Crises of Political and Governmental Rationality

This chapter began with a discussion of the 1843 *Select Committee on Smoke Prevention*. As a key moment in the history of the British State's intervention within atmospheric affairs, the activities and recommendations of the 1843 Select Committee revealed that the genesis of systematic forms of air pollution government in Britain was predicated upon a newly emerging relationship between government and science. It is in this context that this chapter has argued that a history of atmospheric government in Britain has to combine both a history of government with a history of air sciences. In order to explain the particular conceptual postures and methodological techniques deployed within the atmospheric history presented in this volume, this chapter has reflected upon the insights of Michel Foucault's work on the history of governmentality and work within the sociology of scientific knowledge. While Foucault's rich and diverse analysis of the history of governmental reason provides a crucial context within which to position and interpret the nature and intent of British atmospheric government, this chapter argues that notions of governmentality, at least as they are presented in the 1978 lecture series, provide only a partial indication of how one might explore the role of science within the shaping of governmental knowledge production and the associated mentalities of government. This chapter has consequently explored the potential for SSK to provide a methodological context within which it is possible to construct a history of atmospheric government on the basis of its attendant scientific apparatus of knowledge production.

This chapter argues that the notion of *government with science* is suggestive of a set of historical process in and through which certain forms of scientific practice have supported a governmental ethos within the State, and certain governmental desires have fostered the formation of new, and reconsolidation of older, scientific networks of knowledge production. Analysis has consequently shown that much can be gained by combining the insights of governmentality and SSK when constructing a history of air pollution government with science. What both notions of governmentality and SSK have in common is a desire to move beyond the explanations of historical change that are bound to the explanatory force of the State or the influence of great scientists, to reveal the micro-practices of knowledge production and subjugation that run through histories of government and science. In this context, however, it is important to be clear about the normative intent of the histories of government with science that are presented in the chapters that follow. In his reflections on the sociology of scientific history, Hacking observes,

> Philosophers have long made a mummy of science. When they finally unwrapped the cadaver and saw the remnants of an historical process of becoming and discovering, they created for themselves a crisis of rationality (Hacking, 2005 [1983]: 1).

It appears that as philosophers and sociologists have excavated the histories of science a crisis of rationality has emerged that has provided fertile ground for the science wars to flourish.[41] But SSK has not emerged in order to undermine the epistemological or methodological history of science (as Latour himself asks, who is more devoted to the importance and historical significance of science than those who meticulously reconstruct its history in such microscopic detail? Latour, 1999: 1–7). In musing on Hacking's provocative quote, I have become conscious of the fact that it is also possible to interpret Foucauldian inspired histories of government as an attempt to reveal the State as a political institution not of universal power and legitimacy, but of expedient evolution and dilettante practices. Yet as with the SSK, Foucault, and his many acolytes, are not necessarily anti-governmental. Instead through their meticulous histories they have sought to understand the paths taken and rejected by the institutions and personnel of government: they seek to explore the historical contingency of government in order to understand what different governmental futures may look like and how it may be possible to reach such futures (to paraphrase Latour, who is more dedicated to asserting the importance of government than those that seek to meticulously reconstruct their histories?).[42] It is in this context that this book's analysis of atmospheric government with science is, in part, dedicated to considering alternative ways in which it is possible to image developing novel and egalitarian relationships with the air, and the role of governmental and scientific practices in helping to produce the knowledge systems upon which such relationships could be based.

Chapter Three

Science, Sight and the Optics of Air Government

Mr Bloor's Journey: A Day in the Life of a Smoke Observer

On Saturday morning, 31 January 1925, Mr Bloor, of London County Council, commenced a 42-mile round trip of the metropolis. The purpose of Mr Bloor's journey was to observe and record the quality of the city's air on behalf of the Public Control Department of the Council. The following is an excerpt from Mr Bloor's *Smoke Consumption Report* of that day,

> From 10am to 1pm visibility was good, but after that became misty making observation difficult. On the whole as far as smoke was concerned, the observation was disappointing, with hardly any of the numerous shafts emitting smoke during the time I was there. The following are some that were emitting smoke today. The square brick shaft at the Barking Gulford was the worse case of Black Smoke I saw today, close by lay a shaft owned by the Cape Asbestos Co emitting dense black smoke.[1]

There are three aspects to Mr Bloor's report that have particular import for this book's reflections on the governmental science of British air pollution. First, this report indicates the significant amount of time and energy that had to be expended in 1925 to provide reasonably detailed accounts of atmospheric pollution in Britain. The receipts attached to Mr Bloor's report indicate that he had traversed the metropolis by taxi, bus and on foot, and that his endeavours had taken up a whole day.[2] Second, Mr Bloor's report indicates the extreme limitations that smoke observers faced in trying to describe and quantify the nature of air pollution. With the exception of a limited number of gauges and filters (see Chapter Five), in 1925 the monitoring of air pollution was an embodied act that depended on the wit and sensibilities of trained observers. In this context the observation of pollution presupposed a number of ideal conditions that were often not available to the observer: a suitable vantage point, clear weather (with limited wind), not to mention good eyesight. Securing efficient lines-of-sight was not, however, the only restraint associated with the effective monitoring of air pollution by smoke observers. Mr Bloor's observations were obviously limited to visible forms of pollution (although there is some evidence of the olfactory monitoring of certain pollutants by smoke observers).[3] Also, and despite Mr Bloor's apparent willingness to work on what may well have been his day off (31 January 1925 was a Saturday), the monitoring of smoke by human observation could only ever be carried out on a transitory and intermittent basis. The third, and perhaps most surprising, thing to notice about Mr Bloor's report is his evident disappointment at the lack of air pollution filling the air on this winter day. At one level it is important to recognise that the absence of pollution reported by Mr Bloor would not have meant a clear sky: being a winter's day, the air of London would have been filled with the effluvia of domestic chimneys. As the pollution caused by domestic hearths was not subject to government control and regulation, however, it was obviously factored out of the ocular sensitivities of the smoke observer. The disappointment expressed by Mr Bloor does, however, serve as an important insight into the ways in which systems and subjects of government require an object of governance to exist, however it is constructed, in order to legitimate their continued existence and function.

This chapter utilises Mr Bloor's 1925 trip as a fulcrum upon which to explore the systems of government that emerged in Britain around the visual recording and observation of air pollution. As we will see, the observational collection of air pollution data by both amateur and professional practitioners had been going on in Britain for a long time before Mr Bloor made his journey. Smoke observation had been carried out by local smoke abatement societies, factory inspectors and even police authorities throughout much of the second half of the nineteenth century in British cities. This chapter is interested, however, in the emergence of the professionally trained smoke observer as a figure of atmospheric governmentality, and how this figure embodied a response to the limitations of previous optic regimes of pollution observation and ongoing tensions within the emerging governmental science of air pollution monitoring. Much has already been written on the role of sight within the constitution of the modern ordering systems of government and science (see Crary, 1992; Edney, 1999; Foucault 2003a [1963], 2003b [1966]; Haraway, 1991; Rose, 1992; Scott, 1998). Little has, however, been said about the particular ocular challenges that air presents to visual registers of government. By reflecting on the records of nuisance inspectors, medical officers, factory inspectors, smoke observers and police officers this chapter considers how new regimes of atmospheric observation facilitated an embryonic governmental science of air pollution, but an ultimately compromised system of atmospheric governmentality.

Modernity, Sight and the Calibration of the Observing Eye

In a sense the observation of air pollution in Britain has been going on as long as visible forms of atmospheric pollution have been produced. From the concerns over the burning of sea-coal in thirteenth-century London, to John Evelyn's dismay at the presence of smoke in the seventeenth-century Royal Court (see Chapter One), being able to see air pollution has been one of the primary stimuli in attempting to tackle its worst effects. Although atmospheric sight remained an issue in British air government until well into the 1920s, debates concerning the effective visualisation of air pollution were at the heart of a series of attempts to forge a new governmental science of air pollution from the 1840s onwards. While a high proportion of the time taken up by the hearings and witness statements of the 1843 Parliamentary *Select Committee on Smoke Prevention* was devoted to discussing the best practical means of air pollution abatement, Members of Parliament were also interested to hear of the first systematic attempts that were being made to visually assess and record the nature and extent of the pollution problem.[4] With the effective deployment of instruments of air measurement to the assessment of air pollution still some 50 years away, it was the field of vision that would provide the first arena within which the formative dynamics of the governmental science of air pollution would be expressed. But, in order to understand the role of visual practices within the consolidation of early modes of air pollution government, it is important to recognise the broader changes that were occurring in the use of sight as a tool of government and science during the nineteenth century.

In his celebrated account of the reconstruction of modern sight, Jonathan Crary develops a radical perspective on how to interpret the changing history of vision (see in particular Crary, 1992: 1–24; see also Crary, 2000). According to Crary, the transcendent basis for objective sight that flowed from the experiments of Boyle, and informed the epistemologies of the Enlightenment, embraced a fundamentally timeless notion of vision. In classical terms, science was synonymous with a form of vision that was set outside of history. During the nineteenth century, however, Crary discerns a rupture with classical paradigms of vision, a rupture that saw the re-entry of sight into the realms of history. Crucially, Crary does not associate the changes in nineteenth-century sight purely with the artistic rejection of renaissance notions of perspective that were typical of modernism, but with a more varied set of socioeconomic forces. Crary thus observes,

[t]he break with classical models of vision in the early nineteenth century was far more than simply a shift in the appearance of images and art works, or in systems of representational conventions. Instead, it was inseparable from a massive reorganisation of knowledge and social practices that modified in myriad ways the productive, cognitive, and desiring capacities of the human subject (Crary, 1992: 3).

Drawing on the work of Deleuze and Guattari, Crary claims that to understand the changing epistemologies associated with the practices of sight it is necessary to position vision not only within the frameworks of aesthetics (art forms) or technology (novel tools), but also within much broader amalgamations of socioeconomic change and knowledge production (ibid.: 8).[5]

So what occurred in the early nineteenth century to change how vision was understood and, perhaps more importantly, practised? According to Crary it was the unfurling socio-cultural and politico-industrial transformations associated with modernity that would recast the nature of sight. While the birth of modernity was clearly a product of Enlightenment science, and associated techno-industrial developments, as a mode of social existence modernity would define a fundamentally new way of being in the world than that envisaged within the Enlightenment.[6] The new valuation of visual experience described by Crary was, in part, a product of the inescapably subjective nature of sight produced by European modernity. Capitalist modernity would have two fundamental impacts on the amalgamations that constituted nineteenth-century vision: (i) the new complexities associated with the swirling developments of urban industrial life saw an intensification in levels of visual interaction with previously unseen processes and events (Crary 1992: 11); and (ii) the rise of industrial capitalism resulted in the elevation of commodity form and fetish to previously unimagined levels (ibid.: 14; see also Barthes, 2000; Berger, 1972). These two processes (one of the intensification of visual opportunity, the other of the deliberate manipulation of sight and sign) would powerfully challenge, if not halt, the quest for objective visual perspective and truth.[7]

The work of both Crary and Foucault reveals how recognition of the subjective nature of vision facilitated, and in part required, the development of new strategies for the (re)objectification of sight (Crary, 1992: 15; Foucault 1991 [1975]). Drawing on Foucault's analysis of modern forms of disciplinary technology and surveillance, the critical insight of Crary's work is to reveal how the emergence of more subjective (or autonomous) regimes of sight did not emancipate vision, but instead saw the relocation of the scientific project of objective inscription from the technological realms of the scientific laboratory, and onto the human body (Crary, 1992: 16). This transition from mechanistic to human sight; from the cyclopean eye of master science to the binocular vision of the body; from *geometric optics* to *physiological vision*, would give rise to a revised scientific interest in the potential for, and limitations of, the human eye.[8] Concern with how to marshal the capacities of the human eye would see the emergence of what Crary describes as a 'technique of overlapping subjection and objectification' whereby,

> [k]nowledge was accumulated about the constitutive role of the body in the apprehension of a visible world, and it rapidly became obvious that efficiency and rationalisation in many areas of human activity depended on information about the capacities of the human eye [...] The widespread preoccupation with the defects of human vision defined evermore precisely the normal, and generated new technologies for imposing a normative vision on the observer (ibid.: 16).

While Crary outlines how new norms of observation emerged within the mass cultures of nineteenth-century Europe, Foucault highlights how these new sciences of vision and observation became part of novel systems of disciplinary surveillance, both within correctional institutions (such as the prison, asylum and clinic) and the wider population (Foucault, 1991 [1975]; 2003b [1966]; 2007 [2004]). According to Foucault, the recalibration of the human eye during the early nineteenth century did not only facilitate the restructuring of economic consumption (through a new psychology of the fetish), but also enabled a new breed of medical, social and governmental observers to emerge, replete with new powers of diagnosis and knowledge gathering.[9]

It is by no means insignificant that the first systematic observation of smoke commenced during the intensive period of visual recalibration described by both Crary and Foucault. In many ways it is helpful to think of smoke as just one of the swirling array of disordered detritus associated with early capitalist modernity in Britain. As we have already discussed in Chapter Two, the key differences between *pollution artisanale* and *pollution industrielle* was that industrial air pollution was of an altogether different scale and scope than its historical predecessor. In early nineteenth-century Britain the increasing scale of air pollution could be discerned in two ways: (i) in the way in which it moved from being a local, point source problem, to being something that afflicted the public spaces of the city at a pan urban scale; and (ii) by virtue of the fact that it started to afflict an increasingly large number of industrial cities beyond London. The rapid spread of air pollution in early industrial Britain meant that, as with so many of the new visual phenomena associated with modernity, it could no longer be the preserve of the isolated and removed gaze of the experimental scientist, working in their metaphorical obscura. The observation of smoke would require a mobile subject who was able to follow the elusive and confusing smog as it spread through the urban sky and weaved its way through the streets and alleys of the metropolis.

The open acceptance of the need for embodied systems of smoke observation was the stimulus for new institutional structures to emerge that would be dedicated to the training, inscription and regulation of the bodies of the observers who would carry out the work of atmospheric surveillance. In order to begin to understand the training of the modern observer's eye in general, and the smoke observer's gaze more specifically, it is helpful to reflect upon Foucault's analysis of the relationship between *seeing* and *knowing*. Foucault explores this relationship through an analysis of the historical evolution of the clinical style of gaze (2003a [1963]). While provisionally outlining the nature of clinical observation, Foucault reflects on a passage from Corvisart's preface to the French translation of Auenbrugger's 1808 *Nouvelle méthode por reconnaître maladies internes de la pointrine*, 'How rare is the accomplished observer who knows how to wait, in the silence of the imagination, in the calm of the mind, and before forming his judgement, the relation of a sense actually being exercised' (ibid.: 32). According to Corvisart, the key to developing an embodied yet scientific way of seeing the world depended on the ability of the observer to allow time for prolonged engagement with the field of vision. Extended visual engagement with the object under scrutiny was seen to be scientifically important for two reasons: (i) it enabled the full range of relevant visible signals to be registered; and (ii) it facilitated an effective dialogue to be established between the mind and the eye. Time did not only ensure effective visual reconnaissance then, but also served to still the mind and silence the imagination (ibid.). The calm mind and hushed imagination were crucial to the scientific observer because they were seen as the precursor to reason – a reason that had been carefully shaped by systematic visual training and experience.[10] Matthew Edney thus recognises that during the nineteenth century, belief was not based on seeing alone, but on a distinctly *guided* form of vision (Edney, 1999: 48). It was the deliberate directing of vision that was vital to the perfection of this ultimately flawed, but nevertheless indispensable, human sense. To be an effective observer during the nineteenth century was to eschew the desultory gaze in order to allow time for vision to be governed. This is precisely what Crary emphasises in his recognition of the dual etymology of the word *observer*. According to Crary, to be an observer is to be both attentive to rules and objects: to observe norms and regulations as well as things (Crary, 1992: 5–6). The effective administration of the attention of scientific and governmental observers consequently enabled the continuation of an Enlightenment project within the frail corporeality of the human body. The reason for the observing gaze was to both perfect the senses and to still the impulse to deductive theory. Nineteenth-century vision was Enlightenment empiricism continued by another means.

In this volume I am primarily concerned with the role of controlled observation within the act of a specific form of atmospheric government.

To this end it is important to consider the impacts that the transformation in the nature of vision had on specifically governmental gazes. In his analysis of the nature of vision associated with the modern State, James Scott reveals how industrial modernity produced an over-abundant field for governmental vision (Scott, 1998). In this context, Scott recognises that in order to *see like a State* it was important to develop not only an attentive, but also a focused form of gaze,

> Certain forms of knowledge and control require a narrowing of the field of vision. The great advantage of such tunnel vision is that it brings in sharp focus an otherwise more complex and unwieldy reality. This very simplification, in turn, makes the phenomenon at the centre of the field of vision more legible and thus more susceptible to careful measurement and to calculation. Combined with similar observations, an overall, aggregate, synoptic view of a selective reality is achieved, making possible a high degree of schematic knowledge, control and manipulation (ibid.: 11).

The work of Scott reveals how good government requires the construction of a very specific relationship between seeing and knowing. While the clinical gaze requires time in order to engage an analytic of scientific interpretation, the governmental gaze appears to require time in order for the observer to disaggregate and simplify the field of vision. It is the acts of simplification and abstraction that enables governmental authorities to know and understand aggregates like populations, territorial resources and national atmospheres (see Chapter Two). As we move through this chapter it will become apparent that there is actually no sharp distinction between the practices associated with the governmental and clinical gaze. The guided gaze of the medical professional would indeed become a paradigm for the State official. Crucial variance does, however, emerge between governmental and clinical ways of seeing when the institutional reasons and geographical location of sight are explored. When interpreting the role of atmospheric observers it is critical not only to position them in relation to prevailing scientific procedures of sight, but also in relation to the governmental institutions and training regimes in which they operated, and the challenging geographical situations that they encountered within their work.

While Scott's account of the governmental gaze is highly instructive, there is a tendency within his analysis to reduce the reason for State vision to the construction of legible (and thus manageable) aggregates alone. Consequently, whether Scott is describing the trained sight of early modern surveyors and forest managers, or the visions of high modernist town planners, the reasons for seeing like a State do not fundamentally change. It is in this context that Foucault's analysis of the history of governmental reason (outlined in Chapter Two of this volume) becomes so important to the analysis of the governmental gaze. If we take Foucault's broad tripartite history of governmental reason: sovereignty, discipline and security (see Chapter Two), what does this mean for analyses of governmental vision? Well, at the simplest of levels, it would suggest that the rationalities informing governmental sight have not been constant over time, but have instead undergone a series of more or less radical transformations. Recognising this requires a constant sensitivity to the ways in which these different reasons for governmental vision historically overlap, inform and undermine each other within the changing ethos of State power. Scott's account of State sight appears to resonate most strongly with the visual rationalities ascribed by Foucault to the governmentalised State, with its complex apparatus of security. The remainder of this chapter explores the attempts that were made in Britain to generate a scientifically inspired atmospheric gaze within governmental institutions; how this desire was infused and informed by overlapping regimes of visual rationality; and the geographical barriers that the modern city presented to such ambitions.

Nuisance Inspectors and the Constitution of the Legislative Gaze: Between Atmospheric Truth and Atmospheric Proof

When examining the historical construction of a scopic regime of air government in Britain it is important to recognise the exceptionally fragmented nature of atmospheric law and regulation that persisted throughout much of the first half of the early nineteenth century. Up until the 1840s the British government had no systematic mechanisms in place to enable it to effectively intervene within atmospheric relations.[11] Even once successive rounds of national legislation on air pollution were passed, it often fell on emerging institutions of local government to implement them. It is in this context that it is necessary to position early developments in the ways of seeing, recording and regulating air pollution within the nascent systems of government that were emerging in municipal authorities and city corporations. City corporations represent one of the most significant forms of governmental response to the various problems associated with modern industrialism in nineteenth-century Britain. City corporations were essentially bodies with official legal standing that had the mandate to collectively control the various aspects of public policy that had previously been delivered by local charitable organisations or the goodwill of private interests. As precursors to the modern system of local government that exists in Britain today, city corporations gradually assumed responsibility for the provision of a range of public improvements and services that the rapid growth of cities necessitated. If we take the policy area of health (perhaps the most directly pertinent in the nineteenth century to questions of atmospheric pollution) as an example prior to the establishment of urban corporations and municipal authorities collective health care was predominantly delivered by voluntarily funded organisations such as general hospitals, workhouse infirmaries and street commissioners. Through new systems of local taxation, city corporations sought to create a system of public responsibility for health policy and to coordinate the previously disparate bodies involved in the delivery of urban medical support.

While the formation of urban corporations would create a new sense of governmental responsibility (not to mention capacity to act) on air pollution, the nature of incorporation also had an important role in shaping the nature of atmospheric vision. If we take, as an example, the case of Birmingham, which is located in the English Midlands, we see an interesting relationship emerging between early governmental sight and acts of municipal incorporation. Birmingham received its charter of incorporation in 1838. As one of Britain's largest and most heavily polluted metropolitan centres, the act of incorporation was a vital moment in the realisation of governmental ambitions in the area. Suddenly, rather than constituting an urban system with multiple administrative districts, incorporation meant that public works could be conceived of and delivered for the first time at a pan-metropolitan scale. In response to the Public Health Act of 1848, the Smoke Nuisance Act of 1853 and the Birmingham Improvement Act of 1851, the Municipal Corporation formed a Borough Inspection Committee in May 1856.[12] This Inspection Committee was established in order to monitor and enforce these Acts of public health legislation. Crucially, in the context of this book, each of these Acts required the regulation, monitoring and abatement of air pollution.

One of the most important actions of Birmingham's Borough Inspection Committee was the appointment of the city's first nuisance inspector. In addition to having a wonderfully enigmatic title, the nuisance inspector was to act as the professional eyes and ears of the Inspection Committee, monitoring and analysing a range of public health issues. Similar types of nuisance (or public health) inspectors were appointed in other large cities such as Leeds, Salford and Liverpool during the time period within which Birmingham commissioned its first inspectoral activities. As arguably the earliest figures to be charged with the systematic governance of atmospheric pollution in Britain, it is worth spending some time outlining the various duties associated with these nuisance inspectors. One of the main roles of Birmingham's first nuisance inspector was a legal one. The nuisance inspector was responsible for bringing information on smoke pollution events to the Inspection Committee, and on their orders, taking legal actions against those deemed to be transgressing the legal limits placed on atmospheric pollution. The following excerpt, for example, is taken from the Birmingham Borough Inspection Minute Book of 3 September 1856,

The Inspector reports that he laid information against Mr Eddleston and Williams, George Street and Mr Richard Dodge, Bull Ring for Smoke Nuisance. They appeared before the Justices on the 30th and were fined 20 Guineas and costs each and the nuisance abated. The Inspector did not lay any information against Mr Phillip Harvies, not thinking it a sufficient case to interfere with.[13]

The various reports presented by the nuisance inspector for Birmingham make it clear that legal precedent heavily conditioned his governmental gaze. Birmingham's nuisance inspector was not trained to see and quantify air pollution in and of itself, but to assess the legal implications of an air pollution event. The legalistic gaze of the nuisance inspector should not, however, be interpreted as a form of governmental sight that assessed whether some absolute legal boundary had been transgressed, but rather as an analysis of whether a pollution event could be converted into a legal prosecution. The assessment of whether a pollution offence was admissible in court was particularly important to the nuisance inspector for two reasons: first, because he was expected to act as the key witness for the prosecution; and second, because his wages were partly determined by the number of successful prosecutions he delivered (to present an unsuccessful case before the justices would not only be a waste of his time, but also financially prohibitive).

In many ways the legally codified eye of the nuisance inspector reflects the continuation of the litigious form of air pollution government that operated within the courts leet system in pre-industrial Britain (see Chapter Two). The very existence of the nuisance inspector, however, meant that a more systematic, and supposedly efficient, application of the law could be achieved than had been possible under the rather ad hoc system of prosecution brought before courts leet. Despite being grounded in the absolute authority of the law, it is wrong to equate the gaze of nuisance inspectors with a form of sovereign sight and control. Unlike the absolutist proclamations on air pollution made by English and British monarchs in the past, the nuisance inspectors were not policing a system of complete smoke abatement. In the case of Birmingham, for example, the 'nuisance' level of air pollution was calibrated under the rubric of the extent to which pollution was *dangerous or injurious to health*.[14] Calculating the potential human health impacts of an air pollution event was undertaken through regular liaison between nuisance inspector and City Inspection Committee, which counted medical professionals among its membership. The inspector and Inspection Committee in Birmingham met every month. At this regularly convened meeting the inspector's record of public health offences was presented and recorded in the Committee's minute book. While the routine opening of the inspector's book provided an opportunity to assess tolerable levels of air pollution, these meetings were primarily concerned with the legal cogency of the atmospheric testimonies presented by the nuisance inspector. The cases of atmospheric pollution that were referred by the City Inspection Committee for prosecution were consequently not so much revelations of atmospheric truth, but more testimonies of atmospheric proof.

Meat, Lodgings and the Sky: From the Generic Gaze to Specialist Air Observation

Controlling urban nuisance and the generic gaze

While the formation of urban corporations and the implementation of the Smoke Nuisance Act of 1853 gave rise to previously unprecedented levels of atmospheric surveillance, the nature of public health institutions in mid-nineteenth century Britain placed severe restrictions on how the atmosphere could be seen. The problem was that by constructing atmospheric pollution as an issue of public health, atmospheric government become part of an increasingly cumbersome and confusing arena of knowledge gathering, regulation and policy development. The discursive construction of public health offences as acts that were *dangerous or injurious to human health* meant that nuisance inspectors were confronted with a bewildering spectrum for governmental observation. Again taking the case of Birmingham's nuisance inspector, we see nuisance offences covering the conditions of lodging houses, the quality and safety of meat being produced in slaughter houses, the provision of new systems of paving and drainage in public spaces, the regulation of cholera and smallpox, and even the development and location of the first generation of public urinals![15] These varied governmental spheres meant that Birmingham's nuisance inspector was expected to combine knowledge of architecture, food production, sanitation, engineering and epidemiology. In the context of such varied responsibilities it is, perhaps, unsurprising that early nuisance inspectors did not necessarily have specialist training in the observation and abatement of air pollution. Air pollution was just one, albeit crucial, area of the new governmental field that was defined by the term 'public health'. Perhaps the clearest example of the varied nature of nuisance inspection in the industrial city I have found comes from a summary of inspection activities compiled by the Chief Sanitary Inspector of Glasgow in 1887. In his Eighteenth Annual Report (a report which admittedly reflects the gradual accumulation of inspectoral responsibilities over 40 years) he details more than 360,000 acts of inspection spread over 28 areas that range from a lack of light to the availability of handrails on stairs (see Table 3.1).

Table 3.1 Summary of Glasgow Sanitary Inspector's Eighteenth Annual Report, year ending 31 December 1887

Nature of inspections	Number of inspections conducted
Accumulation of garbage	1,731
Apartments with insufficient light	14
Ashpits – out of repair or poorly located	1,830
Bad smells in houses	202
Building where animals were inappropriately kept	73
Corners used as urinals	1,244
Dangerous chimney cans on roof tops	1
Dead animal matter under floor	16
Defective windows	56
Water supply from cistern in water closet	174
Drains, soilpipes, branches	4,943
External walls of dwellings	4,245
Floors of house or water closet	1,033
Handrails and treads on stairs	228
House damp	79
Broken jawboxes water closets or traps	1,984
Lobbies requiring light/ventilation	90
No ashpit, privy or water closet in accommodation	18
Roofs of houses	18
Pipes or gutters of tenements out of repair	1,405
Smoky chimneys	139
Steam or noxious waste discharged from sewers	2
Walls, ceilings out of repair	188
Water cisterns foul or uncovered	73
Water supply damaged	1,732
Windows of staircases	43
Reports of waste of water	558
Nuisance and infectious diseases	361,114

Source: G.Cit.Arch DTC.14.2(6)

To gain a more detailed sense of the generic governmental gaze that was required by early nuisance inspectors it is instructive to consider the summary of nuisance offences that were brought before the justices by Birmingham's nuisance inspector in September 1857 (see Table 3.2). Table 3.2 reveals two interesting things about the inspection of nuisances. First, it is clear that in the pursuit of some 191 prosecutions in one 12-month period that the governance of public health placed a great stress

Table 3.2 Summary of legal prosecutions brought by Birmingham's nuisance inspector between September 1856 and September 1857

Nuisances and offences against:	Total	Male	Female	Discharged male	Discharged female	Convicted male	Convicted female	Fined
Health	46	44	2	11	1	33	1	34
Smoke Act	**11**	**11**				**11**		**11**
Common Lodgings Act	24	18	6	2		16	2	22
Sale surrounding food	87	85	2			85	1	40
Other sanitary offences	23	22	1	3		19	1	20

Source: B.Cit.Arch BCC/AR-74

on the time of Birmingham's nuisance inspector. Second, the relatively small number of (admittedly successful) prosecutions brought for air pollution offences reveals the relatively limited amount of time that was, at this point, being devoted to atmospheric observation. While it is obviously dangerous to equate legal proceedings directly with levels of observations, it is clear that in constituting only 5.8% of yearly public health cases heard in Birmingham in 1856/7, that air pollution did not feature prominently in the nuisance inspector's gaze. This situation should hardly be surprising. The urgent need to address the threats associated with cholera, smallpox and other epidemic threats was high on any mid-nineteenth-century inspector's mind. At a more practical level, the actual opportunities to observe smoke offences were constrained by time. The observation of air pollution required far more time to be spent in the field (waiting for an offence to happen) than, say, inspecting a lodging house or public drain. Yet time was something that early nuisance inspectors clearly did not have as they attempted to glimpse the sky while moving between slaughter-house and sewer, lodging tenement and committee meeting, and between factory and courthouse. It is in this context that it seems likely that the 11 prosecutions for smoke offences recorded in Table 3.2 were instances of extreme public nuisance brought to the attention of the nuisance inspector by members of the public and not as a result of any systematic observation of the city's skies.

The strain that these varied responsibilities of inspection and legal testimony placed on early nuisance inspectors can be discerned in the minutes of the meetings held between Birmingham's inspectors and the City Inspection Committee.[16] In a letter to the Inspection Committee, dated the 14 August 1866, the Birmingham borough analyst (a later name for the nuisance inspector) Alfred Hill expressed his resentment at the amount of unpaid time he had to spend in the courthouse pursuing public health prosecutions,

Dear Sir, with reference to my Bill of Charges of which you spoke to me I beg to say that for some time after my appointment as Borough Analyst I used to receive a Guinea for inspecting the meat, and another for giving evidence upon it, and that when I undertook to perform the numerous duties, sanitary and chemical, mentioned in the conditions of my appointment, including the gratuitous inspection of suspected meat, attendance to give evidence in courts of justice was not included.

I am frequently compelled to wait in court until 2 o'clock, thus losing half a day for which I think the fee of a Guinea is little enough, and it does not seem to be an equitable arrangement that my remuneration should be dependent on the accident of the prosecution being successful; my time and services are of equal value whether the case be gained or not.[17]

Despite now having the new, if still wonderfully vague, title of borough analyst, it appears that Alfred Hill was beginning to feel the strain of his numerous, and often unnecessary, inspection duties and relatively limited remuneration. Significantly, Alfred Hill was a trained medical doctor who was based at Sydenham College, Birmingham. While his medical training would obviously have been useful for many aspects of his work, it would not have provided the specialist knowledge required for the effective monitoring and abatement of air pollution.

Re-observing smoke and making time and space for the specialist gaze

During the 1860s and early 1870s a series of events conspired to fortify the capacity of nuisance and public health inspection committees to record and control air pollution. First, in large cities like Birmingham the ability to bring prosecutions for smoke offences was extended from official analysts and inspectors to include any two physicians or surgeons (or one physician and one surgeon) who observed a pollution event.[18] The use of itinerant smoke inspectors such as doctors and surgeons was further consolidated in large cities such as Manchester where the local police force were deployed as extra eyes on the street (Mosley, 2001: 136–41).[19] Indeed, up until the Public Health Act (London) of 1891 the police authority held primary responsibility for monitoring and prosecuting air pollution offences in the capital (Thorsheim, 2006: 114). Interestingly, when the notion of deploying police offices to assist with the observation or air pollution in Birmingham was made in 1877 the borough analysts opposed the idea. The argument against the deployment of police officers is an instructive one,

> Your sub-committee have also considered the practicability of employing the police in making smoke observations and they are of the opinion that it is impracticable as a length of time must be taken during the emissions of smoke from manufacture's chimneys, and it is absolutely necessary for the inspector to acquaint manufactures when he has committed [an offence] and ascertain the causes of the emission of smoke, and to prove when the case is heard before the justices that it is practicable and possible to consume the smoke.[20]

It appears that when it came to smoke inspection a careful balance had to be struck between the capacities of observation and the ability to provoke and encourage reform. The job of the air pollution inspector, as we will see, was one not just of removed observation, but also of active communion with the perpetrators of air pollution.

The need to forge an intimate working relation between factory owners and nuisance inspectors was heightened in 1863 with the passing of the

Alkali Act.[21] The Alkali Act, and its subsequent amendments, were designed to control the production of hydrochloric acid vapours (produced in the manufacture of alkalis) and various sulphur compounds that were released during the production of heavy chemicals (Thorsheim, 2006: 113). These pollutants were a danger to human health and had caused significant harm to the agricultural lands surrounding large cities (Brimblecombe, 1987: 136–7). At one level the Alkali Act provided a fresh cadre of bureaucrats who were dedicated to a new aspect of atmospheric observation.[22] These bureaucrats formed the nationally constituted Alkali Inspectorate and were overseen by a Chief Alkali Inspector, Robert Angus Smith. At another level, however, the Alkali Act also resulted in much confusion over the role of the different components of the British State within atmospheric inspection. Although the Alkali Inspectorate had legislative responsibility for regulating the sulphur compounds that were produced in the manufacture of chemicals, it was not responsible for the often-greater quantities of sulphur emitted during the processes of combustion. The sulphur produced during the processes of combustion, and contained within smoke, was the rigorously guarded responsibility of fledgling municipal authorities. Thorsheim claims that Birmingham's Inspection Committee resolutely defended its control over smoke abatement in the city against the desires of the Alkali Inspectorate to gain greater control over the regulation of different sources of pollution (2006: 113). While defending its right to regulate smoke pollution, Birmingham's Inspection Committee also gradually assumed more responsibility for different aspects of chemical inspection. With the Alkali Inspectorate being underfunded and understaffed (see Jones, 2007: 111–42), local borough authorities gradually took it upon themselves to regulate different aspects of the Alkali Act at a local level. While taking responsibility for different aspects of the Alkali Act brought with it the promise of greater funding and support for inspection committees, it also carried a new set of inspectoral demands.

The new pressures that the Alkali Act brought to local nuisance inspectors is revealed in the following report delivered by Oliver Pemberton to the Birmingham Borough Inspection Committee in June 1866,

> Sir, In accordance with instructions I received I have visited and carefully inspected the chemical works of James Armitage and Sons situated in Love Lane [...] with the view to reporting on the extent to which gases injurious to health [are emitted] in the processes carried out by the firm. Armitage and Sons are manufactures of nitric and sulphuric acid and of salt ammoniate [...]
>
> Sulphureated hydrogen is likely to escape at two stages when the supply [of] crude ammoniacal liquor is at the landing of the canal – the other when the gas liquor is saturated with hydrochloric acid. In regard to [the] second, I am

of the opinion that no escape of any importance now takes place. I cannot, however, think that this was formerly the case. During the last fortnight Armitage and Sons have arranged that the sulphureated hydrogen given off is at once conveyed into a flue where it is burned and finally permitted to escape in a harmless form via a Chimney 120 ft high.[23]

The need for highly specialised understanding of the various chemical and mechanical processes that caused air pollution, combined with a desire to establish strong personal bonds of reform between inspector and factory owner, placed new pressures on atmospheric inspectors following 1863. These pressures would, in the long term, be crucial to the emergence of a more expert breed of atmospheric inspectors in Britain.

While the Alkali Act of 1863 was an important moment in the move towards a more specialist atmospheric gaze, the relative success of the Act meant that it was another, far more recalcitrant, form of air pollution that would be the target for this new regime of governmental sight: namely smoke.[24] While acid-based forms of air pollution would prove susceptible to governmental regulation, the lack of technological solutions to the production of smoke (not to mention the lack of fuel-substitute alternatives to coal) meant that it become the primary focus of atmospheric observers during the later years of the nineteenth century. Both the 1866 and 1875 Public Health Acts contained important proclamations on the regulation of smoke pollution. Although the 1872 Public Health Act did not address the problems of smoke directly, its requirement that large urban corporations amalgamate the various tasks associated with public health into new *Urban Sanitary Authorities*, and appoint associated medical officers, suddenly meant that more resources were made available for public health inspections generally (and atmospheric inspection more specifically).[25]

The creation of the larger and more powerful Sanitary Authorities, which replaced Borough Inspection Committees, created a new institutional context for atmospheric observation in mid-nineteenth-century Britain.[26] Returning to the example of Birmingham we see the establishment of a specialist smoke subcommittee and dedicated smoke inspectors who were to report to the Sanitary Authority on smoke pollution in the metropolis.[27] The birth of specialised smoke inspectors had two important implications for atmospheric governance in the 1870s. First, it enabled the employment of inspectors who had a greater understanding of the nature of smoke and the causes of atmospheric pollution. In this context, it is significant that many early smoke observers were either coal officers or trained meteorologists (see below). Second, it meant that inspectors had much more time to devote to the patient observation of smoke. To paraphrase Crary (1992), the specialist smoke observer who emerged in the 1870s embodied an attempt to facilitate a managed form of atmospheric attention within the distracting nuisances of the modern city. The new levels of attention that could now be devoted to the inspection of smoke nuisances meant that the disciplinary nature of the legalistic atmospheric gaze could be greatly enhanced. Suddenly there was a figure devoted to uncovering and prosecuting smoke offences who was not burdened by the demands of other inspectorial duties. Perhaps more importantly, however, the expert smoke observer facilitated a further change in the nature of atmospheric government: an extension of the disciplinary gaze into the realms of security and governmentality. The specialisation of smoke observation did not only mean that the emissions of more factories could be policed, but also that atmospheric inspectors could spend much longer periods of time learning how to identify different forms and intensities of smoke pollution and to unlock the principles behind the production and movement of smoke within the atmosphere.

The extension of the disciplinary gaze of the smoke inspector into the more abstract concerns of atmospheric security can be discerned in Birmingham almost immediately after the 1872 Public Health Act. It was at this time that the Inspection Committee started to consider how to redeploy its newfound power and capacities into a more systematic analysis of air pollution. It was in this context that in 1873 the city's Inspection Committee's minutes read, '[R]esolved, that the inspector be instructed to place one of his assistants in some locality where he can watch the smoke of some six or seven chimneys at once, for twelve consecutive days and report the results to this committee'.[28] Following this resolution Mr Bolton, of the Smoke Nuisance Subcommittee, took up a suitable vantage point on Birmingham's Broad Street (in the western sector of the city) to observe a collection of factory chimneys.[29] Mr Bolton devoted up to eight and half hours a day between 20 March and 1 April 1873 observing smoke pollution in this vicinity. The results of Mr Bolton's observational endeavours are recorded in Table 3.3. There are three very important points to note within this seemingly innocuous table. The first is the amount of time that Mr Bolton was able to devote to the observation of atmospheric pollution. In all Mr Bolton spent 77½ hours, over the course of 12 days, inspecting the skies over Broad Street. In addition to undoubtedly giving Mr Bolton a sore neck, such an intensity of observation would have been unthinkable within the time demands of the all-purpose nuisance inspector's day.

The second thing of note within Mr Bolton's record of observations is that it provides a reliable temporal record of air pollution. The recording of pollution events in relation to time marks the first move away from simply disciplining smoke offences and towards trying to understand the underlying logic and nature of air pollution itself. Suddenly the changing nature of the observation of air pollution enabled a series of questions to be asked about its nature. At what times of day is atmospheric pollution greatest? What impacts do certain weather conditions have on the nature of air pollution?

Table 3.3 Summary of Mr Bolton's smoke observations from Broad Street, Birmingham (20 March–1 April 1873)

Date	Hours on duty [time spans]	Hours foggy	Hours clear	Smoke observed	Average smoke per hour
20 March	**6.5** [10–1/2–5.30]	Nil	6.5	1hr 57min	18min
21 March	**8** [8.30–1/9–12.30]	Nil	7.5	2hr 40.5min	22min
22 March	**3.5** [9–12.30]	Nil	3.5	1hr 6.5min	19min
24 March	**8.5** [9–5.30]	4	4.5	1hr 18.5min	17.5min
25 March	**8.5** [8.30–1/2–6]	2	6.5	3hr	27.75min
26 March	**8** [9–5]	1	7	2hr 32min	22min
27 March	**8** [9–1/2–6]	2.5	6	2hr 4min	26.5min
28 March	**8.5** [9–5.30]	4–6	4.5	1hr 18.5min	17.5min
29 March	**4** [9–1]	2.5	2.5	1hr 42min	13.5min
31 March	**7** [9–1/2–5]	Nil	7	4hr 42min	40.25min
1 April	**7** [9–1/2–5]	Nil	7	3hr 2min	26min

Source: B.Cit.Arch BCC/AR-78 – Minute 3756

Is there a cycle of urban air pollution? The ability to observe pollution on these terms would facilitate a fundamental shift in governmental thinking towards air pollution. The shift in governmental rationality that it is possible to discern in Mr Broad's records is from law and discipline and towards security (see Chapter Two). The practices of observation displayed by Mr Broad and Birmingham's Inspection Committee in the 1870s reflect a desire to *work within the reality* of air pollution, and to understand how it is connected to other natural and socioeconomic systems (see Foucault 2007 [2004]: 47). Understanding the reality of air pollution would enable governmental authorities to assess when it was socially necessary, but also economically possible, for atmospheric government to be operationalised.

The third and final dimension of note within Mr Bolton's chronicle of smoke observation is its recording and calculation of average smoke emissions per hour. The ability to be able to calculate average smoke emissions per hour meant that for the first time, in Birmingham at least, a set of atmospheric data was produced that could be recorded and then compared with other measured observations of pollution carried out in different locations and over differing periods of time. This type of calibrated data embodies the modes of numerical abstraction that we now classically associate with statistics.[30] While statistics can take a variety of forms, it is clear that the types of atmospheric knowledge that were starting to be produced in places like Birmingham in the 1870s were amenable to a nascent governmental programme of pollution data gathering and comparison that was beginning to emerge in Britain. The governmental importance of calibrated statistical data stems from the fact that they enabled Inspection Committees, Sanitary Authorities and National Boards of Health to think about air pollution and its abatement in new ways. Suddenly smoke was not something that was to be regulated exclusively at the factory chimney, but could also be imagined – in relation to the reduction of air pollution averages at least – at the scale of the metropolitan district, city, region and even nation.

The emerging desire to understand and thus govern the reality of air pollution in nineteenth-century Birmingham can been seen in the initiatives of the Borough Sanitary Committee following Mr Broad's provisional survey. By 1877 it is possible to discern a desire in the Committee not only to record air pollution events over long periods of time, but also to vary the diurnal timing of observations. Concerned that factory owners were using the cover of the sleeping city to emit their unwanted atmospheric effluent, the Birmingham Sanitary Committee ordered its smoke observers to rise early in the morning in order to commence a new series of dawn surveys.[31] The Report of the Sub-Committee on Smoke reflects that there was little reward for the inspectors' extra efforts,

> [t]he sub-committee report that they have sent out the smoke inspectors as soon as it was daylight in the morning and the inspectors have reported that there was no more smoke made then than at other times of day, but that they had reason to suppose that smoke was made at 6 and 7 o'clock in the morning and your sub-committee have directed that as the mornings get lighter the inspectors shall go out earlier and inspect then.[32]

Despite the new resources that existed to support smoke observers in 1870s' Britain, this minute reveals that even their trained gaze could not penetrate the opacity of metropolitan night. As with so much illicit activity in the nineteenth-century city, it appears likely that significant amounts of air pollution were transmitted under the concealment of the night sky.

In addition to increasing the temporal scope of smoke observation there was a parallel move by many urban Sanitary Authorities to enlarge the spatial range of atmospheric monitoring.[33] Consequently, in addition to encouraging smoke observers to gaze for longer periods of time over a collection of chimneys, borough authorities encouraged the replication of sustained atmospheric observation in simultaneous urban locations. The earliest evidence I have been able to find of systematic, pan-urban surveys of air pollution comes from Manchester. In the Minutes of Evidence submitted to the government's *Select Committee on Smoke Prevention* of 1843 there is a detailed record of an urban-wide smoke observation programme that was carried out by the Manchester Police Commission.[34] It was Colonel Sir

Charles Shaw, the Chief Commissioner of Police, who ordered Manchester's smoke observation programme to be conducted in the early months of 1841. The pan-urban scale of the Manchester survey was organised by the use of Police Division Districts as a way of spatially ordering observations. The smoke survey took place between 27 January and 1 February with synchronous observations being made of different chimneys between the hours of 2pm and 4pm (see Table 3.4).

It is clear from the documents submitted to the Select Committee on Smoke Prevention that the Manchester survey was inhibited by the fact that on many days smoke pollution and weather conditions were so bad that it obscured the factories that were being observed and made it impossible to attribute smoke nuisance to any one polluter in particular (notice the entry of 'obscured' for Chimney C in Table 3.4).[35] Second, and related to this point, while the Manchester survey had an unprecedented geographical scope, it could still only focus on 53 designated chimneys out of 461 potential sites. Consequently, while offering the potential for a more governmental form of gaze, the efforts of Manchester's Police Authority were clearly circumscribed by the limited time and resources that were available to the Commission.

Despite the disciplinary mentalities of the Manchester smoke survey it is apparent that its use of geographically dispersed, but temporally synchronised observations, would provide a paradigm for the new forms of atmospheric observations that started to emerge in the 1870s. In Birmingham the Smoke Nuisance Sub-Committee ordered increasingly widespread surveys of atmospheric pollution.[36] These surveys would contribute to a greater appreciation of the operation of urban pollution systems at various times of the day and year, and in relation to different climatic systems, working practices and economic cycles. There was, of course, a price to be paid for extending the parameters of air pollution observation in both time and space: this was the cost of employing more highly trained and disciplined smoke observers. This was a cost that some city authorities were able to meet more effectively than others.

Despite the new expectations surrounding atmospheric governance in the 1870s, evidence reveals that the personnel devoted to smoke inspection during this decade were distributed in a highly uneven way. In February 1875 there were 43 sanitary inspectors (including two smoke inspectors) operating in the city of Manchester (with a population of 335,339); 22 inspectors operating in Leeds; and 16 in the city of Bristol (Birmingham Sanitary Committee, 1875).[37] Despite being the largest city outside London, Birmingham (with its population of 360,892) had only 12 inspectors operating in 1875, and of these only one employed in a full-time capacity to monitor smoke (ibid.). While the level of inspectorial capacity was obviously limited throughout British cities, the uneven governmental capacities that

Table 3.4 Day 1 observations of smoke pollution from chimneys A–C in Police District A (27 January 1841)

At 10-minute intervals	A			B			C		
	Smoke (min)	Little (min)	None (min)	Smoke (min)	Little (min)	None (min)	Smoke (min)	Little (min)	None (min)
2.00–2.10	–	10	–	10	–	–	10	–	–
2.10–2.20	–	10	–	4	6	–		OBSCURED	
2.20–2.30	–	10	–	–	10	–	–	–	10
2.30–2.40	5	5	–	3	7	–	–	7	3
2.40–2.50	10	–	–	10	–	–	–	4	6
2.50–3.00	4	6	–	7	3	–	10	–	–
3.00–3.10	7	3	–	5	5	–	7	3	–
3.10–3.20	–	10	–	7	3	–	6	4	–
3.20–3.30	4	6	–	5	5	–	10	–	–
3.30–3.40	10	–	–	–	8	2	8	2	–
3.40–3.50	3	7	–	–	5	5	–	10	–
3.50–4.00	6	4	–	8	2	–	8	2	–
2 hours	49	71	–	59	54	7	49	42	19

Source: HC.PP.1843(583)–App.6, pp. 202–8

existed between cities were a source of great consternation in cities like Birmingham. In 1875 Birmingham had one sanitary inspector for every 30,000 inhabitants; this compared unfavourably with Manchester that could boast an inspector for every 8000 residents (ibid.). Furthermore, it was estimated that Birmingham's solitary, full-time smoke inspector was expected to cover a staggering 190 miles of streets in the completion of his monitoring duties. Such a situation led Birmingham's Sanitary Committee to conclude that,

> Sanitary inspection in Birmingham, as at presently carried out, in no sense amounts to sanitary supervision, for the staff is so limited that its capacity is exhausted in dealing with nuisances which are allowed to become intolerable before they are either brought to the notice of the inspectors or are discovered by them. It is the desire of your Committee that we should be in a position to maintain a constant supervision [...] but this change of system will entail a vast increase of work, which will be further augmented by the more minute character of the proposed inspection.[38]

The tension between supervision and inspection, between the minutiae of personal conduct management and the generalities of governmentality, would be a recurring theme for smoke observers throughout the remainder of the nineteenth century. At this point in time, however, Birmingham's Sanitary Committee's protestation led to the award of extra funding and resources to the Committee; a pattern that would be repeated throughout the industrial towns and cities of Britain. Significantly, a part of the extra funding that was directed into inspectorial activities in Birmingham came from central government's *Local Government Board*. As part of this funding agreement the Birmingham Sanitary Committee was expected to furnish the Local Government Board with data on smoke (and other public) nuisances.[39] This, and other local agreements like it, effectively embodied the beginnings of the nationalisation of air pollution knowledge collection and represented an important shift in the level of reality at which pollution was beginning to be conceived and acted upon.

Between Science and Supervision: Coding the Eye and *sub rosa* Observations

The new governmental capacity generated for atmospheric surveillance during the 1870s had two basic consequences. First, it was inevitably connected to the rapid expansion of observatory activity. The new intensity of atmospheric surveillance is captured well in the following report of the Smoke Sub-Committee to the Sanitary Committee of Birmingham City Council,

The Smoke Sub-Committee reports that during the four months ended December 31st 1876 [...] 2188 observations of chimneys have been made, 235 manufactures whose chimneys emitted smoke from 1–10 minutes in the hour have been cautioned; 85 have been reported where smoke was emitted from 10 to 30 minutes in the hour; 86 have been summoned, and 85 have been convicted in penalties amounting to £45. 12,6 and costs amounting to £35. 14,0.[40]

Being able to conduct 2188 observations of air pollution over the course of four months would have been unimaginable under the generic sanitary responsibilities of nuisance inspectors. In addition to increasing the volume of air inspections, however, the new institutional structures put in place by Sanitary Authorities during the 1870s facilitated the training of a new breed of atmospheric professionals.

Second, in the last two decades of the nineteenth century it is possible to discern the increasing professionalisation of air pollution monitoring. In addition to having dedicated smoke inspectors, the sanitary committees increasingly sought to employ smoke observers who had a specialist (and preferably scientific) background in issues pertinent to air pollution abatement.[41] The types of professionally trained personnel who were deemed suitable air inspectors provide an interesting insight into the different governmental rationalities that were informing Sanitary Authorities at the turn of the twentieth century. On the one hand borough councils had a penchant for employing local coal officers. Coal officers provided two clear advantages when it came to the local governance of smoke pollution. At one level they had an eye that was carefully attuned to the subtle variations that existed between different forms of smoke pollution. At another, more analytical level, through their experience of working with coal they had a good understanding of how to read the shade, density and extent of smoke in order to provide a diagnosis of combustion efficiency. The connection that coal officers could make between smoke type and combustion process was seen as vital in relation to offering advice and supervision to furnace operators and factory owners on how to improve the efficiency of their coal-burning procedures. Figure 3.1 is a diagrammatic cross-section of a furnace drawn by a smoke inspector employed by London's Public Control Department. Such diagrams were used to visually express the problems in furnace design that restricted air circulation and exacerbate smoke pollution.

In some ways it is possible to see the necessary balance between the flow of clean air through a mechanically efficient furnace as becoming a metaphor for the circulation of fresh air through an economically efficient city. Understanding the movement of air through the city was, of course, a much more difficult task than the forms of furnace analysis conducted by coal officers. It is in this context that the choice of the second brand of trained

Figure 3.1   Air motion and the supervision of the furnace
Source: L.Met.Arch LCC/PC/Gen/1/33 – Smoke Observation Report

smoke observer was important. Records show that in addition to coal officers, metropolitan authorities also deemed inspectors with meteorological training as suitable persons for smoke observation.[42] While lacking in the skills needed for the optical analysis of smoke, meteorologists held other abilities that were relevant to the scientific analysis and governing of the atmosphere. At one level they were experienced in the use of observational techniques and technologies for atmospheric surveillance (for greater detail see Chapter Five). At a second, and perhaps more important, level, meteorologists were skilled in interpreting the physical climate through which pollution passes and in understanding the atmospheric conditions under which air pollution is either attenuated or exacerbated. With their knowledge of the mechanics of cloud formation, pressure systems and atmospheric currents meteorologists promised a form of atmospheric government that was removed from the factory and furnace, and instead operated at the level of urban pollution systems. If coal officers could interpret the movement of air and smoke though the furnace, meteorologists promised an analysis of how and why smoke moved through the whole urban landscape, and perhaps more importantly still, the construction of pollution forecasts and predictions (see Chapter Seven). The involvement of meteorologists within the emerging governmental science of air pollution monitoring was, of course, just one moment within a much longer historical entanglement of the meteorological sciences and the study of air pollution in Britain.[43]

The combined use of coal officers and meteorologists to monitor air pollution in nineteenth-century Britain does, however, expose another important tension in the rationalities associated with atmospheric government: the recurring tension between atmospheric supervision and atmospheric surveillance. While the expertise of coal officers was obviously used to supervise those involved in combustion activities, meteorologists were employed in order to adopt a more removed form of scientific overview. While notions of supervision and surveillance were not always incompatible, they were far from complementary. As we move through this volume it will become clear that the relationship between surveillance and supervision has been a constant source of tension in the evolution of a governmental science of air pollution.

The professionalising of air pollution monitoring in Britain during the late nineteenth century was undergirded by a series of new administrative devices that were designed to assist with smoke observation. In London, for example, the Public Control Department designed a series of *Smoke Consumption Report* forms upon which smoke inspectors could more accurately and consistently record their daily observations (see Figure 3.2a). This, and other forms like it, provided space for the recording of the precise date, timing and duration of observations, and for the first time facilitated the classification of different types of smoke. In the form shown in Figure 3.2b, for example, we see categories for 'black smoke' and 'light smoke'; while the space provided for 'other comments' was used by the observer to verbally articulate the types of smoke plumes that she/he observed on any particular day. As a vague and unscientifically specified designation for suspended particulate matter it was difficult, even for highly trained observers, to provide effective accounts of the great variety of smoke events that they encountered. The ability to classify different smoke types was, however, important for two reasons. First, many of the existing acts of local and national legislation focused exclusively on the prosecution of black smoke (see Thorsheim, 2006: 114). Second, the classificatory grid provided smoke observers with the ability to record air pollution events even if they were not going to formally prosecute the cases that they observed. One of the virtues of this form was that it provided a mobile technology through which the information collected by different observers could be compared and, if necessary, aggregated into larger data sets.

It is clear that by the late nineteenth and early twentieth centuries institutional changes in local government, new levels of resource provision for

(a)

(b)

Figure 3.2 London County Council (a) smoke consumption form and (b) smoke diagram
Source: L.Met.Arch LCC/PC/Gen/1/33 – Smoke Observation Report

sanitary committees, and nuanced administrative procedures all coalesced to create a new generation of smoke observers. In many ways this new breed of professionally trained atmospheric observer personally embodies the transition that Foucault charts in liberal government from the art to the science of governmental intervention: a threshold between purported subjectivity and objectivity. Yet despite the careful coding of these new observers' eyes, and the assiduous management of their attention through observational devices such as charts and diagrams, it is clear that atmospheric observers were unable to cross the threshold identified by Foucault. One of my favourite examples of this conundrum comes from the account of a London County Council observer, who on 28 April 1899 recounted,

> I had posted myself in the morning on the Westminster side of the river in order to escape observation, but shortly after 10am the two watchmen employed by the firms came over and stood a short distance off evidently to let me see that I had been recognised, and subsequently the owner [of Lambeth Potteries] came to me on Lambeth Bridge [...] he wished me to represent to the Council that he was doing everything in his power to prevent nuisance.[44]

Unlike the ideal of the removed and objective scientist, the smoke observer was always located within the socio-physical landscape of the city. Sometimes this landscape made observation physically difficult (in relation to atmospheric conditions, or the distance that needed to be covered to make inspections). At other times the cityscape facilitated all kinds of subjective encounters between the observer and the observed. As our smoke inspector stands on the banks of the River Thames – betrayed by his familiarity, the observer becoming quite literally the observed – Lambeth Bridge becomes the threshold of government with science: a space between objectivity and subjectivity; between representation and intervention; between science and supervision.

Conclusion: On the Changing Nature of the Atmospheric Gaze

This chapter has revealed that something quite fundamental changed in the nature of government-based atmospheric observation in Britain during the 1870s. At one level this change was connected to the visual capacities of pollution observers. Suddenly freed of their generic responsibilities for public health enforcement and nuisance abatement, Sanitary Authorities were able to foster and tutor a new breed of dedicated atmospheric observers. These observers often had specialist training in how to read smoke and the skies, but perhaps more importantly, were given more time to dedicate to observing urban atmospheres. At another level, however, the change in atmospheric observation that began in the 1870s was not so much about the capacities of governmental sight, but the changing reasons for atmospheric government these new regimes of ocular governance facilitated. Suddenly with more resources, and more observers, the intention of atmospheric government was able to change from one devoted primarily to disciplinary enforcement to a governmentality concerned with the nature of air pollution itself. It is not that the disciplinary role of smoke observers disappeared (indeed with more observers on the streets, sanitary committee minutes show that from the 1870s onwards it was possible to bring more cases of air pollution to the courts than ever before), but rather that it was joined by another reason for government. Through the use of large-scale, synchronised observations of the atmosphere it become possible for sanitary authorities to develop an understanding of the broader atmospheric functioning of air pollution and how it was connected to short-term urban cycles and even longer term economic fortunes. It is important to note that this change in the nature of atmospheric governmentality was not part of some unfurling modernist State agenda, but was instead the contingent outcome of a series of struggles over the local funding of government regimes and the gradual establishment of municipal authorities. Atmospheric science did not simply colonise governmental practice: the incorporation of a scientific mentality into air government was the product of political struggles over institutional responsibilities and resources. What this change did mean, however, was that the question of atmospheric government began to shift from a concern with how to police and enforce atmospheric legislation, to an assessment of how effective atmospheric governance was, and of the most effective ways of constructing regimes of air governance in the future.

Throughout this chapter we have seen how the desire to make the government of air pollution in nineteenth-century Britain a more scientific endeavour focused on the constitution of a new regime of vision. The attempt to make the observation of air pollution more scientific was facilitated through a series of interrelated strategies including: the employment of full-time smoke observers; the recruitment of personnel with backgrounds in areas of atmospheric or pollution science; and the use of prolonged periods of controlled atmospheric observation. In addition to these strategies, a series of devices also enabled the more systematic coding and recording of pollution episodes. These varied devices ranged from the nuisance inspector's book that was 'opened' regularly in front of inspection committees, to the smoke consumption forms and smoke diagrams employed by the London Country Council. While justified in relation to the more accurate and scientific collection and corroboration of atmospheric data, I think it is important to see these varied devices and associated rituals of truth as ways of ensuring that atmospheric observers could themselves be observed and monitored. Perhaps the modern governmental observer is not simply defined by the ways in which her or his vision is conditioned and coded, but by the extent to which the practices of observation are themselves laid open to scrutiny and analysis. Despite all of these strategies and devices, it is clear that even by the time of Mr Bloor's journey in 1925 the art of smoke observation had not attained anything approaching a scientific status. What is crucial to recognise, however, is that the failure of atmospheric observation to attain a more scientific standing was not just a product of the subjective limitations of the observer's body, but was also a consequence of the continuing supervisory ethos of British atmospheric government.

Chapter Four

Governing Air Conduct: Exhibition, Examination and the Cultivation of the Atmospheric Self

At the British Empire Exhibition of 1924–5 the Ministry of Health displayed an intriguing model of a miniature Londoner atop a wooden plinth and replete with shiny shoes, chequered trousers, black overcoat and neatly starched collar. What was most striking about this diminutive figure was that he was leaning over with knees bent and arms outstretched carrying an enormous sack upon his back. The sack contained the actual soot deposited, and then collected by a smoke inspector, in London during one solitary minute. The soot had been enclosed in an imitation sack that resembled those normally used to store and transport coal. The figure of the unfortunate Londoner had been miniaturised by the same scale ratio as the sack of soot to a standard sack of coal. This exhibition piece was a powerful representation of the staggering burden of air pollution that was being produced every passing minute in large British cities. Such innovative exhibition pieces were becoming increasingly common features of a new phenomenon sweeping through Britain at the time: the clean air, or smoke abatement, exhibition. In late-nineteenth and early-twentieth-century Britain numerous clean air exhibitions were convened in places like London, Glasgow, Birmingham and Manchester. What connected these various exhibitions, and is embodied so simply within the Ministry of Health's miniature Londoner, was the desire to shift the weight of atmospheric responsibility from governmental institutions onto the shoulders of the public.

Exhibits, like the model utilised by the Ministry of Health, were part of a broader system of atmospheric pedagogy that emerged in Britain during the final quarter of the nineteenth century. This pedagogic movement was connected to the work of smoke inspectors, which was discussed in the previous chapter, in two ways. First they, in part at least, reflected recognition within the nascent clean air movement that the stringent laws that were enforced by smoke observers were not enough to overcome Britain's air pollution problem. This educational movement consequently sought to address atmospheric effluvia not through the forced compliance of the public, but through the arts of moral and economic persuasion.[1] Second, this movement for educational reform provided an outlet for the new forms of knowledge concerning air pollution, which were being produced by the new cadre of atmospheric inspectors emerging in Britain. This innovative movement for atmospheric persuasion was organised by an elaborate network of social reformers and smoke abatement societies, and spread its message through books, pamphlets, newly commissioned journals and instruction manuals. Two of the most important contexts through which this social reform movement operated were the aforementioned clean air exhibition and the college classroom. Much has already been written on the role of exhibition spaces and classroom instruction in the diffusion and consolidation of both governmental power and scientific knowledge (Bennett, 1995; Naylor, 2002; Yanni, 1999), but this chapter will show how the Victorian penchant for exhibition, display and working-class education shaped the diffusion of, and evolving connections between, atmospheric science and government in Britain. The exhibition and lecture hall provided spaces where atmospheric science and government could meet, mix and travel into the public sphere. As places for the mixing of science and politics, exhibition and lecture halls enabled the intentions of governing institutions to gain scientific legitimacy; emerging clean air sciences to receive crucial institutional support; and atmospheric scientists and smoke inspectors alike to share novel techniques for pollution observation.

This chapter's focus on clean air exhibitions and spaces of instruction facilitates an analysis of the relationship between atmospheric government and personal air conduct. This chapter explores how a system of atmospheric conduct emerged in Britain alongside new knowledge regimes about the nature of atmospheric pollution. Through its focus on exhibition and educational instruction, however, this chapter reveals how new systems of personal air conduct were not only encouraged within the formal institutions and spaces of the State, but also by various voluntary organisations, social reform establishments and private corporations. Focusing on the accounts of those who attended smoke abatement exhibitions, and the various course literatures and handbooks that were produced to support the promotion of smokeless technologies and practices, this chapter considers how the government of socio-atmospheric relations braided with discussion of personal health and hygiene, debates over gender relations within the home, domestic science and kitchen design, and the practices and procedures of the workplace. Ultimately, analysis reveals how key scientific discourses of the male and female self were utilised in order to govern the atmospheric individual and nurture a new regime of personalised atmospheric care.

Exhibition Spaces, Gendered Practices and the Atmospheric Responsibilities of the British Home

The exhibition as governmental technology

While we often take the spaces of exhibition associated with public art gallery, museum and commercial stall for granted they are each relatively recent historical phenomena. In his classic account, *The Birth of the Museum*, Bennett argues that it was the staging of the great public exhibitions and world fairs of the nineteenth century (and in particular London's *Great Exhibition* of 1851) that enabled new understandings of the potential and technological capacity of exhibitions to be realised (Bennett, 1995). Furthermore, Bennett claims that far from being innocent acts of public display and beneficent pedagogy, during the nineteenth and twentieth centuries exhibitions become increasingly important sites within the strategic articulation of various forms of power and knowledge (ibid.: 61). In this section I claim that it is no matter of little importance that the emergence of the exhibition, as a new configuration of power and knowledge, corresponded with the rise of widespread public concern, scientific study and political action towards air pollution. Indeed, as I hope to show, the exhibition became a crucial technology within the emerging systems of governmental and scientific study that surrounded air pollution.

Before discussing the role and operational dynamics of clean air exhibitions it is important to reflect upon the connections that exist between exhibitions and governmental and scientific power. To this end Bennett's analysis of the *exhibitionary complex* is a crucial starting point. According to Bennett,

> The institutions comprising 'the exhibitionary complex' [...] were involved in the transfer of objects and bodies from the enclosed private domains in which they had previously been displayed (but to a restricted public) into progressively more open and public areas where, through the representation to which they were subjected, they formed vehicles for inscribing and broadcasting the messages of power [...] (original emphasis) (ibid.: 61).

There are two important things to note in this passage. First is Bennett's use of the term *exhibitionary complex*. Bennett utilises the idea of a complex to reveal that the influence of the exhibition is not limited to the walls of the transitory exhibit, but should be interpreted in relation to the dioramas, galleries, public museums, handbooks, programmes, newspaper reports and catalogues through which the knowledge associated with exhibitions is circulated and supported (ibid.: 60–1). The second point of note is the association that Bennett makes between exhibitions and the *broadcasting of messages of power.* Bennett argues that exhibitions are not only shows of sovereign power and associated forms of great achievement, but also provide a context within which large numbers of subjects could learn to live in ways that were commensurate with prevailing structures of power. Thus returning again to Bennett, he reflects,

> The exhibitionary complex was also a response to the problems of order, but one which worked differently by transforming the problem into one of culture – a question of winning hearts and minds as well as the disciplining and training of bodies [...] through the provision of object lessons in power – the power to command and arrange things and bodies for public display – they sought to allow the people, and *en masse* rather than individually, to know rather than to be known, to become the subjects rather than the objects of knowledge (ibid.: 63).

Work within the history of science has also long acknowledged the dual role of the exhibition and museum within the consolidation and spread of scientific knowledge and power (see for example Naylor, 2002: 500). At one level the careful ordering of exhibition pieces has provided an important, non-laboratory based context for the transfer of elite knowledge and learning between established scientists. At a second level, however, the exhibition has provided an invaluable way of transmitting scientific achievement to the masses through technologies of amusement and entertainment (Morus, 1998: 70–89).[2]

What interests me most about Bennett's analysis of the exhibitionary complex, and is perhaps of most direct relevance to this volume, is his engagement with Foucault. Given that Bennett sees exhibitions as crucial intersections of power and knowledge, it is unsurprising that he is drawn to Foucault's oeuvre. Despite utilising Foucault's analysis of the intersections between power, knowledge and order (particularly in relation to Foucault's *Order of Things* (2003b [1966]) Bennett ultimately sees the exhibition in contradistinction to Foucault's analyses of power. According to Bennett, Foucault's account of the *carceral archipelago* of prisons, clinics and asylums is an analysis of the disciplinary techniques of confinement and reform that are largely marginal within the exhibition (see Foucault, 2003a [1963], 2002 [1961], 1991 [1975]).[3] Bennett reflects,

> This is not to suggest that technologies of surveillance had no place in the exhibitionary complex but rather that their interaction with new forms of spectacle produced a more complex and nuanced set of relations through which power was exercised and relayed to – and, in part, through and by – the populace that the Foucauldian account allows (Bennett, 1995: 61).

For Bennett then the exhibition is more an aspect of the society of spectacle than of the Panopticon; more cultural power than hard-edged disciplinary coercion.[4] It is my contention that if we take a closer look at the full scope of Foucault's oeuvre it is possible to discern a set of perspectives that can offer critical insights into the articulations of power that appear to be expressed in exhibitionary complexes of different kinds.

My desire to connect clean air exhibitions with Foucault's work is not derived from a stubborn aspiration for intellectual consistency. Rather I contend that by conducting a Foucauldian analysis of the exhibitionary complex it becomes possible to see the exhibition not merely as a general expression of cultural power, but as a specific form of governmental technology. To do so, however, it is necessary to move beyond Foucault's earlier analysis of disciplinary power and to consider his work on biopolitics and governmentality, and his later reflections on the care of the self (Foucault, 2007 [2004], 2004 [1997], 1998 [1976], 1992 [1984], 1990 [1984]). As was discussed in Chapter Two, if we consider what Foucault wrote and taught in the second half of the 1970s we see a much more diverse sense of power emerging than is evident in his earlier work. In this later work notions of disciplinary power and the society of the Panopticon are set within a broader sweep of historical power ranging from absolute sovereignty to more liberal forms of security and governmentality (Foucault, 2007 [2004]). At least two key insights emerge from Foucault's later work on power that have import for the exhibition. First, is the realisation that any history of power should not be read as sequential: with perhaps a period of absolutism being followed by a discrete period of disciplinary control; and then by a more nuanced form of cultural power. According to Foucault expressions and technologies of power overlap and interrelate in much more complex ways than this neat historical narrative suggests. This insight is critical for any analysis of clean air exhibitions. As this chapter will show, while clean air exhibitions did embody new expressions of cultural power identified by Bennett, they were also vital points of passage for the technological devices, techniques and strategies that were vital in the formation of intensified regimes of atmospheric surveillance and disciplinary control beyond the exhibition hall (a dimension of the exhibition that Bennett does recognise, but tends to underplay).

The second critical insight of Foucault's later work on power is his desire to understand the connections and tensions between the technologies of the individual (or how individuals come to be governed as part of a society) and the technologies of the self (the means by which we learn to *look after one-self*) (Foucault, 2007 [2004], 1992 [1984]). Bennett's analysis is, of course, keenly aware of the role of exhibitions as devices for reaching large numbers of people at a very personal level. What Bennett's perspective on the exhibition does not afford, however, is an analysis of the role of governmental analytics of existence in shaping the exhibitionary experience. In this chapter I assert that the clean air exhibitionary complex was an expression of a broader apparatus of atmospheric government. This apparatus of government sought to cultivate a sense of atmospheric self-responsibility, but as a very specific expression of power this form of atmospheric governmentality tended to construct a particular type of atmospheric self: a self positioned within the broader governance of socioeconomic and environmental relations. In the clean air exhibition the care of the self thus became enmeshed in the care of a broader population. I thus agree with Bennett when he asserts that exhibitions,

> [s]ought also to allow the people to know and thence to regulate themselves; to become, in seeing themselves from the side of power, both the subjects and the objects of knowledge, knowing power and what power knows, and knowing themselves as (ideally) known by power [...] (Bennett, 1995: 63).

As technologies of government, clean air exhibitions specifically enabled the public to know what nascent systems of atmospheric government and science knew and to thus make atmospheric decisions that were not purely motivated by selfish desire. The question then becomes one primarily concerned with the particular types of atmospheric self that are cultivated within governmental societies, and how these may vary from those produced under systems characterised more by commercial, imperial or disciplinary power. As the following section will illustrate, the governmental self bears the individualistic hallmarks of the free consumer, but a consumer that calibrates action along scientific lines of calculation and in relation to an ethos of collective socio-environmental care.

The birth of the clean air exhibition: 'public laboratories' for pollution abatement

The exhibition of clean air technologies and practices in Britain was pursued through a series of specially convened smoke abatement conferences. During the late nineteenth century, and for much of the first half of the twentieth century, clean air technologies were also displayed and demonstrated at more generic events including inventors' symposia, ideal home exhibitions and the numerous displays that were organised by gas and electrical departments of different metropolitan corporations. Such displays, exhibitions and conferences were often organised by voluntary smoke abatement societies, sanitary reformers and local health alliances.[5] Large clean air exhibitions were often supported and endorsed by government departments and were filled with the exhibits of numerous companies and inventors. It appears that in addition to being conditioned by the general Victorian penchant for education through exhibition, clean air exhibitions were also influenced by two movements of the late nineteenth century in particular. Thorsheim (2006) interprets clean air exhibitions as part of a broader set of events instigated by sanitary reformers in order to promote the improvement of a gamut of socio-environmental problems facing British cities.[6] Clean air exhibitions, with their strong promotion of smokeless electrical appliances, were also shaped by a much longer history of electrical exhibit. Morus describes how during the early nineteenth century spaces of exhibit throughout Britain were converted into galleries for the practical sciences associated with electricity (1998: 70–98). Exhibits of electricity were designed to educate the public about the wondrous potential of this new energy form, and to assuage any lingering uncertainties that people held over its dangers. It is clear that many of the innovative exhibition techniques and styles of scientific display developed to promote the work of electrical scientists and engineers in the early nineteenth century were deployed within the spaces and stalls of later clean air events.

While conditioned by these broader forces of exhibit, the story of clean air exhibitions in Britain really begins on 30 November 1881. It was on this day, in rooms attached to the Royal Albert Hall in Kensington, London, that Britain's first major smoke abatement exhibition was opened. London's Smoke Abatement Exhibition was organised by the Smoke Abatement Committee (which would become the Smoke Abatement Institution).[7] The Smoke Abatement Committee was a broad collection of scientists and social reformers who were concerned with the deleterious effects that air pollution was having on British health and economic vitality. Ernest Hart was Chairman of the Smoke Abatement Committee and W.R.E. Coles was its Honorary Secretary at the time (Coles would eventually be appointed smoke inspector for London by the Home Secretary).[8] While the Smoke Abatement Committee was based in London it saw its advocacy role as extending throughout Britain and to any city that was subject to the menace of smoke.[9]

In order to gain an insight into the form and functioning of the 1881 Smoke Abatement Exhibition it is instructive to reflect upon the account of a correspondent of *The Times* who visited the exhibits,

> Such are the leading features of the smoke abatement exhibition, which is replete with ingenious contrivances for preventing the waste of fuel and the formation of smoke with regard to both domestic and industrial purposes. The claims of utility made by newcomers will, to a certain extent, be met by the committee of investigation, and a number of inventions will go forth to the world with the stamp of approbation upon them (*The Times*, 1881: 11).

Visitors to the exhibition came from an array of different backgrounds and included housewives, servants, engineers, furnace workers and members from a range of different trade unions. Those attending the exhibition were treated to a number of active exhibitions with new cooking ranges and ovens in operation, as well as furnace and boiler demonstrations. An important role of the exhibition was not only to display and demonstrate smokeless and energy efficient devices, but also to judge and assess the different exhibited items. The exhibition's inspection committee, who assessed the relative merits of different inventions on display, presented awards, medals and certificates. The notion of *improvement through competition*[10] has always been an important rationale of exhibitions and shows of various kinds, but in the case of the 1881 Smoke Abatement Exhibition nomination for an award had important commercial implications. A crucial role of the exhibition was to act as a kind of commercial forcing house, in and through which the Smoke Abatement Committee could officially endorse and support devices that they felt would contribute most effectively to the quest of smoke abatement. By bringing inventors, engineers and designers from across Britain together, it was also anticipated that the event would help to develop synergies for the continued refinement of smoke abatement technologies.

Early clean air exhibitions were proclaimed successful in part because of the large numbers of people who attended them. It is reported that the 1881 Smoke Abatement Exhibition drew crowds of approximately 116,000 people, while Manchester's exhibition of the following year attracted 32,000 visitors (Mosley, 2001: 148). On other levels, however, it appears that these early smoke abatement exhibitions were less successful. There was a pervading sense amongst many in attendance that while the events revealed attractive, and at times inexpensive, solutions to air pollution, the displays were too far removed from the realities of everyday domestic and industrial life to be taken seriously. A correspondent for *The Times* consequently reflected,

> How far they [smoke abatement devices] may ultimately be instrumental in helping purify the atmosphere from the nuisance of smoke it is difficult to surmise. To those of a domestic character this especially refers, inasmuch as many of the stoves are dependent for their proper action upon close watching and careful management. This the inventors and assistants can give them while in their hands, but this the ordinary run of users and domestic servants cannot and will not give to them in daily practical use. However, the exhibition can hardly fail to do good in teaching the public that there are conditions under which coal can be burnt without smoke [...] Although but few may succeed in attaining the perfection of domestic firing now to be witnessed at South Kensington, many may be induced to make the attempt, and may attain partial success (*The Times*, 1881: 11).

As demonstrations in the *perfection of domestic firing* it appears that early smoke abatement exhibitions were forms of governmental simulacra: versions of smoke abatement practices that could find no original in the real world of the bustling home and pressured workplace.

Other critiques of London and Manchester's exhibitions focus less on the ability of the exhibits to translate their breakthroughs into the real word of the smoke-laden city, and more on the intentions of those who actually visited the shows. Writing on Manchester's exhibition Mosley observes,

> The working-class attendance at the Manchester Exhibition was good, perhaps making up as many as 20,000 of the 32,000 visitors. But the vast majority almost certainly came to hear the band that played every Sunday afternoon and to view such novelties as Tyndall's Musical Flames, rather than out of real enthusiasm for smoke abatement and the 'cheerless' alternatives to the open hearth (Mosley, 2001: 151).

It would clearly be misleading and inaccurate to suggest that smoke abatement exhibitions initiated some fundamental and large-scale shift in the nature of atmospheric conduct in Britain. I do, however, want to claim that clean air exhibitions provide an insight into the new strategies of commercially supported governmental control that were used to target the British home in the late nineteenth and early twentieth century. While not initiated exclusively from exhibition spaces, it is further claimed that these governmental strategies became part of a broader socio-cultural shift in the practices of domestic maintenance and reproduction (see Figure 4.1). In this section I want to argue that clean air exhibitions embodied a governmental technology that enabled emerging forms of atmospheric government to reach individual citizens en masse. At the same time, exhibitions offered a channel of delivery in and through which the practical technologies needed for self-directed atmospheric government could be promoted and distributed.

Exhibiting technology and ordering the spaces of exposition

With the idea of the exhibition as a technology of government in mind, let us consider the different tactics employed at British clean air exhibitions to promote domestic reform. It is clear from the various accounts and records of clean air exhibitions that significant time and attention were spent ensuring that the design and layout of exhibits not only made technologies visible but also highly desirable. At one level the effective presentation of smokeless technologies was facilitated through the design of individual exhibits. Significant effort and commercial expertise were put into the construction

Figure 4.1   The enlightened home. Photograph of visitors attending the 1910 Glasgow Clean Air
Exhibition
Source: G.Cit.Arch P.175

of many individual exhibits at clean air exhibitions. A report in the *County
Municipal Record* of the first *International Smoke Abatement Exhibition* (held
in London in 1912) reflects on the elaborate design of the Gas Department's
stall at the London Corporation's exhibit,

> The Gas Department of the Corporation exhibits many devices for gas heat-
> ing, lighting and cooking. Their display is housed within a very handsome
> compartment of Doric design, divided into compartments showing kitchen,
> dining room, bath-room, shop window and general show stall. The display
> [also comes with] its gas laundry and cooking demonstrations.[11]

At Glasgow's Smoke Abatement Exhibition of 1910 the *Glasgow Herald*
describes a similarly elaborate domestic reconstruction,

> The model house which the Gas Department of the Corporation have fitted
> up is an epitome of the exhibition. In its rooms are to be found the most
> improved systems of incandescent lighting by James Milne and Son (Limited)
> and M'Innes and M'Lachlan. The gas appliances for heating and cooking,

and laundry are most varied, and for them all is claimed economy, efficiency, convenience, and above all, cleanliness and fumelessness.[12]

In addition to the construction of clean air homes, public and private exhibitors also spent considerable time designing bespoke recreations of smokeless rooms. For example, the *Glasgow Herald* described the commercial production of a highly attractive smokeless dining room at the city's 1912 exhibition,

> The Dining room, which is handsomely panelled in Austrian Oak, is provided with two fireplaces, in one of which a 'St Nicholas Fire' made by Messrs R and A Main and Co, is fitted, and in the other one of the latest gas fires made by Messrs John Wright and Co [...] Not only is the latest form of heating demonstrated in this apartment, but the lighting arrangements are a special feature. There are handsome pendants and candle brackets manufactured by Messers James Milne and Son, while on the mantle piece there are figures carrying small lights.[13]

The ability to fabricate the smokeless home was a crucial objective of clean air exhibitions. Such totalising domestic reconstructions were an important part of the mixing of commercial strategies and government objectives at exhibitions that enabled effective targeting of the domestic sphere (see Figures 4.2 and 4.3).

The promotion of the smokeless home was, in part, based upon the formation of a sharp contrast with the smoke-based domestic scene. The coal-fired home was one of time-consuming energy production, with open fires, stoves and laundry facilities having to be continually attended and supplied with heavy fuel. In addition to being difficult to maintain, the coal-based home was often a cold place with open fires being ineffective at heating large rooms and corridors. Worst of all, perhaps, the coal-fired home was characterised by a sooty atmosphere, dirt and a continuous accumulation of grime. The objective of smoke abatement exhibitions was to construct the smoky home as a kind of pre-scientific, archaic, and even primitive form of human existence, soon to be replaced by the resplendent efficiencies of gas and electricity. Witness this account of the launch of Sheffield's 1909 Smoke Abatement Exhibition,

> Sir Oliver Lodge,[14] declaring the exhibition open, said that the earth was beautiful in the extreme where nobody lived (laughter), but where people lived together in large numbers they had taken no precautions to keep the earth beautiful. [...] A savage could burn coal as we have burnt it in our grates. A better way was to separate coke and gas at the pit, and burn only gas in the house [...] The atmosphere in which we live was not an atmosphere in which our history had been made, and our history now seemed to get smoky and foggy (Laughter) (*The Times*, 1909: 10).

Figure 4.2   The St Mungo's fireplace exhibit at the 1910 Glasgow Clean Air Exhibition with female attendant
Source: G.Cit.Arch P.177

The connection that Sir Oliver Lodge makes between the burning of coal and more 'savage' forms of existence is an important one. A common thread that connects the clean air exhibitions hosted in Britain during the late nineteenth and early twentieth centuries was their desire to position smoke-less technologies within a broader epochal shift. This was not just a shift from coal to gas, the open grate to electricity, inefficient to efficient coal combustion, but between archaic forms of domestic existence, and a new scientifically inspired existential paradigm. It is in this context that the Doric columns of the London Corporation Gas Department's exhibit were as symbolically important as the ideal home it contained. The Doric design was not merely a cosmetic gimmick, it was a direct architectural reference to the Greek scientific rationality that the smokeless home was meant to reflect: it was Pythagoras meets Worcester Bosch; Archimedes in a gas-heated bathtub.

The purpose of clean air exhibitions was more than merely the selling of commercial merchandise, it was about positioning the need for smokeless technology within the biopolitical context of atmospheric reform and

Figure 4.3   A smokeless kitchen exhibit produced by the Corporation's Gas Department at the 1910 Glasgow Clean Air Exhibition
Source: G.Cit.Arch P.183

socio-historic change. In this context, a significant amount of attention was given within exhibitions to the overall layout of the exhibition hall. Through the careful spatial orchestration of stalls, those responsible for clean air exhibitions juxtaposed visions of the smokeless home with vivid imagery of the deleterious effects of air pollution. If we take the 1936 Smoke Abatement Exhibition (held in London's Science Museum and run by the National Smoke Abatement Society) as an example, the spatial design of the exhibition becomes clearer. According to the exhibition handbook and guide, the hall was divided in three main sections: (i) smoke and its consequences; (ii) techniques for the measurement of air pollution; and (iii) the latest methods available for smoke abatement (National Smoke Abatement Society, 1936: 46).[15] A report published in *The Times* illustrates how the first objective of the 1936 exhibition was achieved,

> The exhibition is designed to illustrate three things – the nature and effects of smoke, the methods by which air pollution can be measured, and how the nuisance can be abated. Some of the most striking exhibits illustrate the first

point, and they include a fine series of photographs showing the great smoke-pall over London, the blackened clouds over Epping Forest after a London fog and view of Middlesbrough, Edinburgh, Manchester, and other towns [...] to demonstrate the effect on human health there are specimens of uncontaminated and contaminated lungs: and a diagrammatic arrangement of little black crosses showing how in December, 1930, a month without fog, there were in Manchester 137 deaths from respiratory diseases, whereas in January, 1931, when there were nine days of fog, the number of deaths rose to 592 (*The Times*, 1936b: 11).

Through charts, maps and graphic representations clean air exhibitions attempted not only to promote smokeless technologies, but also to illustrate the often-neglected human costs of air pollution.

It is interesting to reflect on the diverse collection of objects and ephemera brought together in smoke abatement and clean air exhibitions in order to illustrate the diverse consequences of air pollution. Number 12 in the *Catalogue of Exhibits* for the 1936 exhibition was the aforementioned display of human lungs. This collection of lungs was lent by Professor S.L. Cummins and contained,

(a) Lung of child 3 years old, healthy and not yet contaminated by smoke or dust. (b) Typical lung of city dweller, showing distribution of carbonaceous dust under the pleura and in some of the tracheo-bronchial glands, with a little dust in the lung tissues. (c) Lung of coal trimmer with no chest symptoms. Carbonaceous deposits throughout lung tissue. (d) Lung of colliery borer suffering from silicosis. The incombustible impurity of the lung contains 42 per cent of silica (National Smoke Abatement Society, 1936: 47).

This, admittedly gruesome, biological exhibit was clearly intended to invoke a desire for the medical care of the self and others among visitors to the exhibition. Unlike the Gas Department's Doric compartment, this exhibit did not facilitate a movement into a domestic space of the future, but instead enabled a journey inside the human body. As a kind of corporeal technology, these blackened lungs served to raise public consciousness of the hidden consequences of air pollution. To use Foucauldian terminology, the exhibits embodied an anatomical politics that utilised the effects of air pollution on the internal fabric of the cadaver to convey the need for broader biopolitical regimes of reform. As exhibits the human lungs themselves captured the scalar shift Foucault identified between the narrow medical concerns of *anatomo-politics*, and the broader regimes of social health associated with biopolitics (Foucault, 1998 [1976]: 139).

Table 4.1 provides details of the full range of exhibits on the effects of air pollution that were displayed in London in 1936. The majority of these exhibition pieces were on temporary loan to the Conference Committee.[16]

Table 4.1 Catalogue list (with numbers) of exhibits at the 1936 Smoke Abatement Exhibition (London) concerned with 'Smoke and its Effects'

1. A Lancashire cotton town (photograph)
2. The Potteries (photograph)
3. Infra-red photograph of London and its environs
4. Aerial photographs of air pollution by Captain Alfred G. Buckham (FRS)
5. Half-cleaned oil painting
6. Transparencies of smoke abatement photographs
7. Specimens of soot-incrusted brick
8. Photographs of stone and brickwork showing the effects of atmospheric pollution
9a. Photographs showing the effects of smoke pollution on buildings
9b. Micro-slide of smoke-affected sandstone
9c. Natural size photograph showing portions of Portland Stone patera about 130 years old
10. Sermon in stones. A photograph of Kings Street Manchester
11. Sections of weathered limestone
12. Specimens of lungs showing effects of atmospheric pollution
13. Graphic chart showing effects of smoke-fog upon deaths from respiratory diseases
14. Chart showing examples of fatal smoke-fogs
15a. Ultra-violet ray meter
15b. Loss of ultra-violet radiation in Manchester (chart)
16. Specimens showing the effects of atmospheric pollution on vegetation
17. Influence of gaseous pollution in the tarnishing and 'fogging' of non-ferrous metals
18. Influence of sold pollution (disperse particles) on the rusting of iron
19. Influence of gaseous pollution in the formation of green patina on copper
20. The absorption of sulphuric acid from polluted town atmospheres and its influence on the durability of leather
21. Leather bindings from Windsor Castle
22. Records of atmospheric pollution in certain libraries for one year
23. Prize photographs
24. Atmospheric dirt from air-conditioning plant
25. Specimens of various types of grit and dust from furnaces
26. Acid pollution of atmosphere (glass jar)
27. Linen exposed to city atmosphere
28. Fragments of chimney pots

Source: National Smoke Abatement Society, 1936: 46–50

Two further exhibits are, I think, worth particular note. First, Exhibit 16, 'Specimens showing the effects of atmospheric pollution on vegetation'. This exhibit was loaned to the exhibition by the Royal Botanical Gardens at Kew, and came with the following note,

> Apart from the loss of sunlight due to a smoky atmosphere most forms of vegetation suffer directly by the presence of solid impurities and obnoxious gases in the polluted atmospheres of big cities. Plants reared under such conditions do not attain their full growth and luxuriance, while plants reared in a clean atmosphere are visibly affected when introduced into a polluted one. Susceptibility to the various impurities differs for different kind of plants [...] (National Smoke Abatement Society, 1936: 48).

While concerns about the impacts of air pollution on plants often had a biopolitical dimension – particularly in relation to the effects of pollution on agricultural crops and harvests – it is clear that the role of such exhibits was in part designed to instil a desire for an environmentally sensitive governmental rationality (see Chapter Eight).

The second exhibit of particular note in the 1936 exhibition was provided by the Leather Manufactures Research Association (Catalogue number 20). This exhibit was simply entitled 'Leather bindings from Windsor Castle' and came with the following description,

> All three [leather bindings] were bound at the same time (1903) with leather from the same delivery; the first two, which were stored at Buckingham Palace for 10 years, are decayed, and the leather has absorbed 5 per cent. of its weight in sulphuric acid. The third, which had remained at Windsor, has only absorbed 3 per cent. and is much better condition (ibid.: 49).

The accidental experiment that had been unconsciously carried out on the surfaces of these leather bindings served to remind exhibition goers of the impacts of atmospheric pollution on national treasures and artefacts.[17] When placed alongside the eroded fabric of Georgian and Victorian buildings, such exhibits conjured the image of a national heritage that was being gradually lost to the corrosive influences of an industrial atmosphere: a kind of atmospheric re-rendering of national history.

Foucault reminds us that the practices of government should not be simplistically equated with the 'the formulae employed to convince, persuade, and lead men [sic] more or less in spite of themselves'; in other words government is not equivalent to 'politics, pedagogy, or rhetoric.' (2007 [2004]: 165). I state this observation now because I think that it is easy to read the clean air exhibitions and exhibits described in this section as an expression of the ideological State: or the ways in which State influence on the construction of popular belief systems and assumptions is realised in myriad sites beyond the formal institutions of government, including the school, church, museum and clinic. Yet as distinctly governmental forms of technology I want to claim that smoke abatement and clean air exhibitions were not about the construction of some form of false consciousness among the population, but the desire to develop an enhanced atmospheric awareness. At one level it is important to acknowledge that clean air exhibitions deployed the commercial cunning of gas and electrical appliance providers in order to construct a smokeless home that embodied the qualities of both fetish and simulacra. But such aspiration-based marketing was conducted alongside a set of exhibits that were designed not to instil conspicuous consumption, but a desire for the care of the self, community and even nation. It is interesting here to note the role of mobile things like human lungs, plants and even leather bindings in enabling the varied effects of air pollution to travel directly into the urban consciousness. These diverse carriers of government provided a materially immediate and highly visceral context for personal reform. It is not that pedagogy and rhetoric are absent from the exhibitions we have so far considered, but that the direction of such strategies is towards government: towards the construction and maintenance of the *right disposition of collective things*; a disposition within which the air can support industrial progress without threatening bodily well-being, ecological diversity or national heritage.

What marks clean air exhibitions out as technologies of modern government is not, however, just the goal of achieving the right disposition of atmospheric things, but the ways of governing they seem to suggest. According to Foucault what distinguishes modern forms of governmental power from regimes of sovereignty and discipline is that while sovereignty and discipline image the enforcement of other forms of reality, governmentality attempts to work with and within the nature of contemporary reality in order to achieve governmental aims (see Chapters One and Two) (ibid.: 47). Consequently, while sovereign and disciplinary techniques of government may utilise laws and surveillance to control desire, governmentality is predicated upon the cultivation of personal preference not its artificial negation. In classic governmental terms, the *proper disposition of things* is not presented as an onerous imposition on the individual and society, but as a movement towards a more natural state of socioeconomic and human–ecological balance. Within clean air exhibitions it is thus possible to discern attempts to cultivate a governable atmospheric self. This cultivation process was in part based upon the various exhibits that connected the individual exhibition goer to the atmosphere in very immediate ways. Consequently whether it be the internalised relations between atmosphere and self embodied in the lung specimens, or the more abstract connections between self, atmosphere and public health demonstrated through the various charts and diagrams displayed, clean air exhibitions clearly offered an apparatus of connectivity between individuals and the air.

Learning to live differently and the formation of public womanhood

A crucial aspect of British clean air exhibitions was their role within the direct education and instruction of the visiting public. In focusing upon the overt pedagogic functions of clean air exhibitions attention is inevitably drawn to the constitution of a particular form of gendered subject: the modern housewife. A significant portion of the pedagogic functioning of clean air exhibitions was devoted to the training and retraining of women (and schoolgirls) in the use and application of smokeless technologies. It appears that if clean air exhibitions were primarily about facilitating the government of the British household, then the housewife was seen as a crucial subject in the reformulation of this field of existence.

A sense of the gender politics that surrounded and informed British clean air exhibitions can be gleaned from this short except from the *Glasgow Herald* that was published in conjunction with Glasgow's 1912 smoke abatement exhibition,

> The housewife does not pay much attention to what kind of fuel is used for the boiler, being of the opinion that the cheapest, dirtiest and smokiest coal were good enough, hence the pollution of the atmosphere [...] it should not only be borne in mind that they [smokeless technologies] lighten the labour connected with washing, but that were they to become universally used they would materially help the Corporation in its crusade against smoke.[18]

In addition to displaying a surprisingly chauvinistic attitude towards women, this quote reveals the dual logic of the clean air exhibition with regard to female subjectivity. First, clean air exhibitions sought to generate a new sense of public womanhood by raising female consciousness of the role of the domestic space within the generation and abatement of smoke nuisances. Elsewhere I have argued that the moral repositioning of the housewife within the public affairs of the British city was akin to a form of domestic environmentalism (Whitehead, in press). As a distinctly governmental strategy of moral reform, domestic environmentalism sought to use a collective resource – the urban environment – as a basis upon which to transform prevailing codes of female responsibility. This transformation process essentially sought to relocate the locus of household morality from the running of a thrifty home to the creation of a more outward looking and environmentally benign domestic sphere.

Second, and at the same time as they were promoting a more publicly oriented female subject, clean air exhibitions were recruiting housewives to act as vanguards in the use of new coal-based, gas and electrical appliances. This retraining process took a variety of different forms. Lecture halls were put aside for special training and instruction sessions concerning the most efficient uses of smokeless appliances were given.[19] In addition to the more traditional forms of lecture-based instruction, clean air exhibitions also put on a range of demonstrations of working appliances. Here is an account of a laundry demonstration that was held at Glasgow's 1912 exhibition,

> [h]ousewives are also taking an unusual interest in the laundry work, a display which shows washing and wringing machines at work and all kinds of irons suitable for domestic and factory use in operation [...] What a picture appears in the mind's eye of the housewife as she looks at the neat and daintily dressed laundresses at work in the exhibition buildings. They deal with the most delicate fabrics with ease compared with the conditions experienced in many homes.[20]

In addition to laundry displays clean air exhibitions also held regular cooking demonstrations and competitions.[21] These cooking events were designed to enable housewives, house servants and schoolgirls to learn how best to use gas and electrical appliances in the preparation of different foodstuffs and meal formats. Beyond the halls of clean air exhibitions, the retraining of housewives in the use of smokeless technologies was also supported by local women's electrical associations and a range of new courses provided by technical colleges.[22]

What interests me most about the gender dynamics of clean air exhibitions are the apparent tensions that governmental constructions of womanhood at such sites appeared to entail. As I have argued elsewhere, clean air exhibitions embodied a tension common within the governmental orchestration of personal conduct (ibid.). While clean air exhibitions promoted and promised a publicly minded and engaged woman (now ready to step beyond the myopia of household survival and reproduction in order to consider the issues facing the city, and even nation), the intensified retraining of housewives and servants in the use of kitchen, heating and laundry appliances re-inscribed women more firmly than ever in the domestic sphere (Llewellyn, 2004; Walkerdine & Lucy, 1989; Whitehead, in press). This tension appears to be typical of governmental strategies for the cultivation of new modes of conduct. When modes of self-conduct are re-inscribed in relation to governmental projects (as they clearly were in clean air exhibitions), personal practices, habits and customs are inevitably positioned within a broader fabric of public concern. While the reform of personal (atmospheric) conduct within systems of government is undergirded by an extended sense of moral sensibility, the government of conduct tends to re-inscribe the existing positionality of the subject with such intensity that the opportunity for new and lasting forms of civic existence is heavily curtailed. The fact that clean air exhibitions explicitly targeted the social stratum labelled 'housewife' is symptomatic of this governmental tendency. The housewife was seen as the guardian of the home and was thus the most efficient target for the instigation of effective governmental reform in the domestic sphere. Here government exploits a repressive and gendered reality to serve governmental ends without necessarily redressing the action of repression itself. 'Good government' was then not about female emancipation (although it would eventually become so), but about working with the power relations of the home to achieve the desired disposition of public and private things (see Butler, 1990).

Surveillance and the clean air exhibition

The final aspect of British clean air exhibitions I want to consider in this chapter relates to their role in supporting the emerging science of air pollution observation and monitoring. As discussed in the previous chapter, the advancement of air pollution monitoring procedures in the 1870s and 1880s corresponded with the staging of the first clean air exhibitions in Britain. Here smoke abatement and clean air exhibitions provided an important context within which smoke observers could share their new techniques and the technologies for atmospheric surveillance. New ideas concerning the accurate measurement and recording of air pollution were pooled through the lectures and more informal discussions held in the theatres and corridors of exhibitions.[23] Beyond this system of knowledge sharing, however, clean air exhibitions were important sites for the display of novel technological devices and analytical procedures designed and patented for atmospheric monitoring.

If we return to the catalogue list for London's 1936 Smoke Abatement Exhibition we get a sense of the wide range of atmospheric monitoring displays that were typically on show at such events. Table 4.2 provides a summary of the catalogue list of exhibits dedicated to the measurement of smoke and air pollution at the 1936 exhibition. These devices were loaned to the exhibition by groups including the Manchester Literary and Philosophical Society (Number 29), the London Science Museum (Number 30), Messrs Radiovisor Parent, Ltd, the Fuel Research Station, the Department of Scientific and Industrial Research, and the Meteorological Office. The specific exhibits took three main forms: (i) displays of air monitoring equipment; (ii) illustrations of methods of analysis of air pollution and results gained from different devices; (iii) maps revealing the current levels of atmospheric surveillance being conducted throughout Britain. In relation to the first two forms of exhibit it is clear that part of the function of clean air exhibitions was to promote best practice among the smoke observers and abatement enforcers who were present at the exhibition. Just

Table 4.2 Catalogue list (with numbers) of exhibits at the 1936 Smoke Abatement Exhibition (London) concerned with the 'Measurement of Smoke and Pollution'

29. Thompson air pollution recorder
30. Measuring the density of smoke in a factory chimney (model)
31. Smoke alarm equipment

Exhibitions lent by the Fuel Research Station:
32. Determination of the opacity of a column of smoke (tinted glass instrument)
33. Relationship between the optical properties of a column of smoke and the weight of solids carried by it per unit volume
34. Determination of the weight of solids in the smoke
35. Types of smoke produced by the domestic fire
36. Typical results (physical properties of pollution produced by burning different types of household coal) (graph)
37. Smoke emission from a domestic fire (graph)
38. The combustion of coal
39. Atmospheric pollution per ton of coal, fuel oil or gas consumed

Exhibits lent by the Department of Scientific and Industrial Research:
40. The measurement of atmospheric pollution (map of monitoring stations)
41. Deposit gauge (device and map where device is deployed)
42. Typical samples of the results of analysis of the contents of a deposit gauge
43. Typical results with deposit gauges
44. Selection of site for the deposit gauge
45. The automatic filter
46. Observations with automatic filter
47. Suspended sooty impurity (winter 1934–5)
48. Dust sampling instruments
49. The Owens jet duster counter
50. Thermal precipitator
51. Apparatus for measuring dust and sulphur dioxide
52. Lead peroxide gauge
53. Directional lead peroxide gauge

Exhibitions lent by Meteorological Office:
54. Diagram showing London's loss of sunshine owing to smoke
55. Diagram showing the progress of smoke abatement between 1881 and 1836 revealed by London's sunshine
56. Three diagrams showing visibility at a close network of stations in England and Wales on 13 February 1936, and the effect of the smoke of congested areas on the visibility
57. Two illustrations showing the method of production of high fog and its effects in London

Source: National Smoke Abatement Society, 1936: 50–6

as the displays of the coal-free home and the deleterious effects of air pollution were designed to cultivate a new form of atmospheric citizen within the general public, the celebration of the most successful and effective atmospheric monitoring devices and methods of air analysis were intended to nurture an improved an ever-more reflexive air pollution observer.

The third type of exhibit was clearly conceived, however, to encourage the spread of air pollution monitoring and surveillance throughout the country. The maps illustrating the local authorities that were cooperating in emerging national regimes of monitoring were designed to instil a spirit of cooperation among representatives of local government districts who were not currently conducting systematic air surveillance. Recognising the role of clean air exhibitions in promoting and supporting air monitoring and surveillance is important because it brings into question Bennett's assertion that exhibition spaces reflected a kind of post-surveillance and post-disciplinary power. While clean air exhibitions were clearly involved in the cultivation of forms of power that targeted the everyday cultural practices of citizens, they also appear to have supported more traditional techniques of disciplinary gaze. This said, it would be erroneous to assume that the clean air exhibitions were displaying monitoring equipment whose primary function was disciplinary. In supporting the distribution and use of standard monitoring equipment and techniques, clean air exhibitions were points of passage for a new governmental science of atmospheric security. They embodied, if you like, parastatal sites through which new scientific techniques of atmospheric knowledge production could travel. As sites that mixed cultural forms of power, with disciplinary desires and security concerns, perhaps exhibitionary spaces are most effectively thought of as theatres for multiple powers and rationalities of government. While these different modalities of power and government may operate along very different trajectories they all appear to be commensurate with an exhibitionary model of influence.

Recasting Workplace Conduct: Boilers and the Stoker as Scientist

In this final section I want to move beyond the exhibition hall in order to consider the coordination of atmospheric conduct among a different social group: those workers who dedicated their careers to the operation and maintenance of industrial boilers and furnaces of different kinds. Of course, boiler and furnace operators were, in part, the targets of clean air exhibitions. Many clean air exhibitions (particularly the earlier ones) had large sections dedicated to the latest in boiler and furnace technology.[24] However, with early legal regulations surrounding the emission of smoke and air pollution from industrial premises, governmental intervention within industrial systems of combustion could take a more direct form than that offered by clean air exhibitions. In this section I want to consider how, despite the more draconian mechanisms of government that were at the disposal of the local and national state, the reform of industrial combustion practices came to involve the cultivation of a new subject type: a subject type that was this time more scientist than citizen.

Chapter Three described the role played by nuisance and smoke inspectors in attempting to reform the working practices of boiler operators and stokers. Through the close bonds that were forged between inspector, factory owners and specific workers it was felt that effective gains in the efficiency of combustion practices could be gradually produced. If the smoke inspectors did act as early atmospheric mentors to stokers and boilers workers, it is clear that their role in cultivating new forms of industrial conduct was severely restricted by available time and resources. With the supervision and training of combustion engineers constituting only an incidental aspect of their already overburdened work, it became increasingly clear that the instigation of new industrial modes of conduct could not be left to the supervision of inspectors alone.

Retraining the stoker: from labourer to scientist

During the early part of the twentieth century a number of official courses in stoking and boiler maintenance were instigated throughout the country.[25] Technical schools normally convened these training courses with input and guidance being provided by local smoke abatement societies and committees. While often convened on an ad hoc basis, there were many attempts to standardise the training received by stokers and boiler operators. In 1929, for example, Ernest Dickinson – a technical school lecturer – produced a handbook entitled *Successful Stoking and Smoke Abatement: A Manual for Boiler Attendants* (Dickinson, 1929). In this volume Dickinson attempted to set out a standard curriculum covering the basic sciences of combustion and best technical procedures to deploy when using different forms of boiler. In his manual for boiler attendants Dickinson provides us with an insight into the difficulties of reforming workplace practices,

> At these classes the fireman is taught simply but certainly that there is more to the art of stoking than just throwing coal on the fire. Of course every fireman commences these classes with the idea that if he 'doesn't know how to stoke, then nobody else does.' It universally rules that every fireman is ready to say 'If I can't keep steam nobody else can.' After a few attendances at the

classes he begins to feel that he doesn't know so much, and when the question of certificates arises he feels he would rather be excused (ibid.: 51).

It appears that the professional pride of boiler attendants presented barriers to the newly emerging training programmes that were being put in place for them.

In contrast to the visions of civic duty and public mindedness that pervaded clean air exhibitions, it is interesting to note that a very different strategy was developed for constructing a governable male atmospheric subject in a workplace setting. Within various courses and guidance notes for boiler operators in early-twentieth-century Britain it is possible to discern a discourse that sought to recast their labours in distinctly scientific terms. Listen to H.G. Clinch, for example, in this excerpt from his 1923 *The Smoke Inspector's Handbook*,

> The aim of the stoker should be to extract the greatest possible amount of heat from the fuel used, and it is a great pity that stoking should be regarded as an unskilled trade. The prevailing idea that a strong arm is the only qualification needed in a stoker is quite incorrect [...] Stoking should be classed as skilled labour, and paid for as such, and it is probable that if some method of teaching these men the principles underlying their work were universally adopted, the return to the employers would be surprising, whilst the community generally would reap the benefit in the numerous advantages of a cleaner atmosphere (Clinch, 1923: 31).

While recourse is clearly made here to the wider corporate and civic benefits of more efficient systems of combustion, it is the professional repositioning of the stoker as scientist that is most significant. Clinch indicates the need to transform the idea and self-image of the stoker as an unthinking strong arm and unskilled trade, into a more attentive and reflective character. He goes on,

> The stoker will be something more than a man who slaves with the shovel, with sweat pouring off him, and not caring, or, indeed, having time to care, whether he is wasting coal or not. He will be more or less a scientist, and his life will be considerably more pleasant than it is at present, but he will have to thoroughly understand the principles of efficient combustion (ibid.: 32).

Clinch's reconstruction of the stoker as an attentive scientist, who understands his boiler and the intricacies of its operations, is emblematic of the broader scientific ethos surrounding the re-coordination of workplace conduct being developed in Britain at the time. It is also a further example of the fusing of science and government in the control of air pollution in Britain.[26]

The vision of stoker as scientist served a number of purposes. First, it provided a context within which boiler attendants of various kinds could rationalise and legitimate the extra time they would have to spend being retrained in a profession they already held expertise in. While boiler attendants may have resented and actively resisted didactic forms of compulsory training, the promise of becoming a scientist appeared to assure a new regime of respect for their work and a potential increase in their own personal level of job satisfaction.[27] The ideal of the stoker scientist was, however, more than mere persuasive rhetoric. Local inspection authorities and factory owners wanted boiler attendants who could regulate the efficiency of combustion with scientific precision and understanding. The science of boiler use and maintenance not only promised a new workplace identity for the stoker, but also cleaner airs for the city and cost savings for their corporate employers. The retraining of workers in the techniques of combustion provides an interesting insight into the nature of *government with science* discussed in Chapter Two. While at one level it is clear that the science of combustion engineering benefited from its association with various certifying institutions of atmospheric government, this is not simply a case of government-supported science. Through the adoption of a scientific discourse in the reform of the boiler room it is also clear that the label of science enabled forms of atmospheric government to travel into the notoriously recalcitrant spheres of the workplace.

Ernest Dickinson's *Manual for Boiler Attendants* offers an insight into the form and nature of the courses that were put in place to create purportedly more scientific boiler attendants. Dickinson's handbook covers a range of topics from the basic chemistry of combustion to the technical operation and working of particular boiler types. At the end of his handbook Dickinson describes a rudimentary system for certifying and further professionalising boiler maintenance. According to Dickinson what was needed was a formal examination procedure that would reliably test what boiler operators had actually learned in class. It was important that this test contained a formal written examination (in order to test the theoretical components of the course) and a more practical test carried out with an actual boiler. The examination paper that Dickinson devised, and which was to be used after the first year of his boiler attendants' course, indicates the nature of the scientific training he envisaged (see Table 4.3). The examination paper required the stoker to understand the chemical mechanisms by which combustion was achieved, what may prevent efficient combustion and how firing can be improved. The paper sought to test the stoker's ability to isolate and 'read' his boiler as a scientist would stand back and assess an ongoing experiment. Ultimately though, it appears that the production of stoker examinations, and associated modes of certification, were not primarily concerned with testing boiler attendants. These examinations were designed

Table 4.3   Specimen examination paper for first-year stoker as devised by Ernest Dickinson

SPECIMEN EXAMINATION PAPER

Three questions to be answered from Sections A and B. Number of marks per question is shown.

Section A

1. Of what does coal consist?	(16)
2. What takes place when a boiler fire is stoked?	(18)
3. How does the quantity of air admitted to a boiler affect the fire?	(18)
4. Where does the greatest transfer of heat occur in the boiler?	(10)
5. What can be done to assist the transfer of heat to the water?	(10)

Section B

6. What is a good guide to the condition of a boiler fire?	(10)
7. How can smoke be avoided? Describe a good method of firing.	(16)
8. What effect has bad brickwork on the working of a boiler?	(10)
9. Describe a good way of 'cleaning out,' and mention the points to be noted.	(16)
10. State a number of things which require special care and attention if smoke is to be avoided.	(16)

Source: Dickinson, 1929: 51–2

to instil a sense of self-reflection and the monitoring of personal conduct on behalf of the stokers themselves.

Conclusions – On the Nature of Atmospheric Government and Personal Conduct

All forms of atmospheric government are ultimately attempts to coordinate personal forms of conduct. Whether it be in terms of the practices and procedures associated with atmospheric surveillance, the domestic practices of the housewife, or the workplace activities of boiler attendants, governing the atmosphere always starts by governing people. Governing the atmosphere is not, however, simply about changing the ways key individuals relate to the atmosphere at a personal level. Atmospheric government (as opposed to other forms of atmospheric power) involves the construction of socio-economically viable sets of relations between a population and an atmosphere. Furthermore, attempting to govern aggregate entities like populations and atmospheres requires particular styles and techniques of government. These systems of government have to be able to connect individuals to the socio-atmospheric consequences of their actions, while coordinating conduct across a large range of actors.

This chapter has revealed that atmospheric government involves two interrelated process: (i) the recasting of personal conduct in relation to the re-coordination of behaviour across key social groups (the urban dweller, employer, housewife and boiler attendant); (ii) the repositioning of the subject in relation to the broader environmental consequences of atmospheric pollution. The processes and techniques of government are vital in respect of both processes: in the first instance, facilitating the coordination of personal conduct through collective education, training and certification; and in the second instance, confronting individuals with the collective atmospheric consequences of their personal choices through the compilation and aggregation of knowledge of environmental change and its varied effects.

The focus of this chapter on the spaces of the exhibition and classroom has revealed that the government of atmospheric conduct can be pursued in very different ways. The voluntary spaces of the exhibition hall were able to use the technologies of exhibit and mass marketing as a way of targeting the largely unregulated domestic sphere. On the other hand, existing legislation on the production of air pollution from commercial premises meant that the spaces of tuition and examination that targeted boiler attendants could take a more compulsory and disciplinary tone. While applying different styles of governmental power, however, it is clear that both the clean air exhibition and the retraining of stokers employed the same governmental message: the need to move from an uncivilised and unthinking age of air pollution into a more scientific time of reason and control. In the domestic sphere this transformation was based on a move from the 'savage' burning of coal to a new science of domestic engineering and electricity. In the boiler room this shift was predicated upon the reconstruction of the stoker from 'brute labourer' into astute chemist and engineer. To these ends it is important to notice how *atmospheric government with science* not only involves the reconstruction of the air as a combined object of political and scientific analysis, but also sees the combined powers of State and sciences recasting social identities.

Chapter Five

Instrumentation and the Sites of Atmospheric Monitoring

What follows is an account of perhaps the most important technological device in the history of Britain's air pollution monitoring,

> The gauge [...] consisted of a vessel of 4 square feet catchment area, having a conical bottom so arranged that all rain falling in the gauge vessel was collected in bottles placed underneath [...] the gauge vessel was of enamelled sheet iron, square in plan, and supported in a frame on four legs; it was surrounded by a cage of wire netting open at the top to prevent birds from contaminating the project.[1]

This device was deployed by *The Lancet* (working under the auspices of the Coal Smoke Abatement Society) to conduct a survey of air pollution in London during 1910 and 1911 (see Figure 5.1). Operating out of *The Lancet's* own laboratories, and supervised by S.A. Vasey, this study was one of the earliest – and certainly most significant – instrument-based studies of urban air pollution conducted in Britain.[2] What is, for me at least, most arresting about this account of the *rain deposit gauge* is the degree of precision and inherent materiality that is conveyed within this passage. As will become clear, the production of precise, durable and replicable instruments for atmospheric measurement became a central concern of the clean air movement, and associated branches of atmospheric government, in the early part of the twentieth century.

The Lancet's rain deposit gauge became an important tool within the development of spatially extended systems of instrument-based air pollution monitoring in Britain. Rather than merely seeing such appliances as instruments of atmospheric government, however, this chapter explores their active role within the production of certain ways of knowing the air and as key conduits connecting the capacities of atmospheric science and

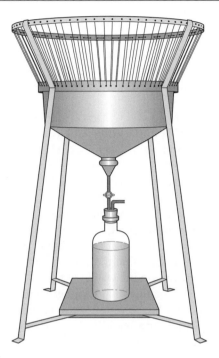

Figure 5.1   The deposit gauge

the desires of air government. The agency of atmospheric measurement devices, such as the rain deposit gauge, in part derives from the fact that they embody a paradox. The prime advantages associated with the use of instruments within air pollution monitoring in early-twentieth-century Britain were fourfold: (i) they were not, on the whole, dependent on the degree to which air pollution was manifest as a visible phenomena (as smoke observers were) (see Chapter Three); (ii) they could provide a constant, 24-hour record of air pollution (something nuisance inspectors did not have the time to provide); (iii) they ostensibly removed the element of subjectivity associated with the imperfect and corruptible eye of the atmospheric observer (see Chapter Three); and (iv) they could be moved and distributed widely in space. Yet for these advantages to be realised devices had to be fashioned with a degree of precision and durability that could not always be guaranteed. In order for monitoring instruments to provide an objective, reliable and spatially extensive record of air pollution they had to be manufactured to a level of exactitude that was hard to maintain and monitor. At the same time, the operation of such gauges was challenged by their need to survive the corrosive elements they were actually measuring and the regular interventions of unpredictable intruders (hence the enamelled sheet iron and wire caging of the deposit gauge). The point is that the fabrication of glass, enamel, sheet iron, rubber and wire that collectively constituted the early instruments of atmospheric surveillance in Britain presented important material considerations and constraints to emerging systems of air government.

Much has already been said and written on the role of instruments within the history of science (see Golinski, 2006; Jankovic, 2000; Shapin & Schaffer, 1985; Schaffer, 1992), and to a lesser extent government (Braun, 2000; Edney, 1999; Latour, 1998 [1984]).[3] At a methodological level, the rising historical interest in instrumentation appears to reflect the more general analytical prioritisation of the micro-practices and local constitution of both science and government expressed within work on the sociology of scientific knowledge and Foucauldian-inspired governmentalities (see Chapter Two). In keeping with the broader intent of this volume, the primary aim of this chapter is to consider the impact that the construction of a combined history of *government with science* has on the way in which we understand the role of instruments within the production of atmospheric knowledge. Existing work on the history of science has given significant attention to the role of instruments (present as basic tools, mechanical devices or digitised systems) within the laboratory-based production of experimental facts (Latour & Woolgar, 1986 [1979]; Shapin, 1996: 96–100). But the role of instruments as what Shapin terms *fact-making machines* becomes problematised when explored in the context of governmental knowledge production (1988). As Foucault reveals, the cognitive desires associated with modern forms of governmentality have given rise to the construction of ever more elaborate networks of knowledge gathering that aspires to comprehend the entirety of human existence (2007 [2004]: 95–6). The construction of such governmental registers of knowledge necessitates the expanded relocation of the instruments of science from the contrived spaces of the experimental laboratory into the varied public spaces of a political territory over extended periods of time. Of course, as *immutable mobiles* instruments have always played a crucial role in the circulatory travels of science (Latour, 1998 [1994]). In the context of governmental sciences, however, instruments are not only involved in the movement of scientific knowledge production capacities between the enclosed spaces of one laboratory to another, but are responsible for the movement of science through a much less predictable set of physical and socio-cultural terrains. It is these processes that bring into focus the various practical problems and barriers that exist to the formation of what Latour refers to as a *durability and extension of interaction* (2006: 72).

It would be disingenuous to suggest that the history of instrumental science has been solely, or even predominantly, focused on laboratory-based practices. Valuable insights into the operational dynamics of the governmental sciences of atmospheric pollution can be gleaned from recent studies of the field-based sciences of natural history, pedology, geography and meteorology (Latour, 1999: 24–79; Naylor, 2006, 2002; Withers, 2001). Perhaps the most important insight of such studies is not the historical insights they provide, but the crucial links between geography and science they reveal. At one level the geographies of sciences associated with work on field-based environmental research reveals the place-based compromises of the sciences of nature: or the ways in which particular landscapes and local environmental conditions influence the instruments it is possible for scientists to deploy in their studies. At a second level, such work has exposed the often arbitrary conditions of landscape and morphology that determine the precise locations where scientific instruments can and cannot be deployed within the study of environmental systems. Finally, studies of the history of field-based sciences have also exposed the complex circulatory dynamics that surround the geographical spread of scientific instruments. In this context, research has revealed that the circulation of scientific instrumentation does not simply involve the even geographical spread of approved devices from certified centres of calculation and calibration to more distant localities, but entails a much more dynamic set of dialogues and exchanges between overlapping scientific communities. In excavating the rise of the instrument-based science and government of atmospheric pollution, this chapter considers each of these important geographical issues. Ultimately, however, analysis shows that the role of instruments in supporting the circulation of science is not necessarily commensurate or compatible with their role in undergirding the knowledge-gathering apparatus of government.

It is important to realise that this chapter does not follow from Chapter Four in direct historical sequence. Instead it reflects upon a series of processes that emerged out of the events and tensions described in both Chapters Three and Four. In relation to Chapter Three this chapter considers how instruments for measuring atmospheric pollution were developed as a direct consequence of the barriers that smoke inspectors experienced when trying to accurately observe and register air pollution. This chapter is connected to Chapter Four in two important ways. First, the move towards the systematic use of instruments within the measurement of pollution was in large part inspired and directed by the events of the clean air exhibitions described in the previous chapter. Second, this chapter seeks to move beyond the focus on the relationship between government and the human subject analysed in Chapter Four in order to critically reflect upon the importance of nonhuman things (both animate and inanimate) within the practices of government.

Technological Deployments, 'Normal Air' and the Appointment of the Committee for the Investigation of Atmospheric Pollution

It was not until after 1912 that an integrated system of instrument-based air pollution surveillance began in Britain. It was in 1912 that the first International Smoke Abatement Conference was held. This conference was convened by the Coal Smoke Abatement Society and, as with the clean air exhibitions described in the previous chapter, sought to promote various strategies for air pollution abatement. One of the key recommendations of the 1912 conference was the establishment of a systematic framework for air pollution monitoring in Britain. It was in this context that the 1912 conference appointed the *Committee for the Investigation of Atmospheric Pollution* (hereafter CIAP). This Committee was composed of a range of people with an interest in air pollution and/or an expertise in the measurement of atmospheric phenomena (see Table 5.1). The Committee drew together representatives from the Meteorological Office, the Coal Smoke Abatement Society and Smoke Abatement League, medical and air purification officers from municipal boroughs, and even a member of the Hamburg Smoke Abatement Society. As the Chair and Honorary Secretary respectively, Sir William Napier Shaw and John Switzer Owens would play a crucial role in the technological developments and political struggles that surrounded the spread of the instruments of air pollution monitoring throughout the UK (see also Chapter Six). While Napier Shaw is, perhaps, best known for his work in meteorology (where he developed the millibar and tephigram), in partnership with John Switzer Owen he became a key figure in the development of a governmental science of air pollution in Britain.

The CIAP met for the first time on 21 June 1912. A sense of the purpose of the Committee can be gained from this excerpt from its first annual report,

> It was felt the time had arrived for the various efforts [of atmospheric pollution monitoring] to be coordinated so that the information could be put in the form of results which could be reasonably comparable for the same place, from month to month, from season to season, from year to year, and for different places for the same periods. It is only in this way that the information can be effectively utilised as an index of present effort and a guide to future action.[4]

In essence, the CIAP embodied a critical shift in the rationality of pollution government in Britain. In Chapter Three we saw how, during the nineteenth century, pollution observation was transformed from a litigious practice of

Table 5.1   Original members of the Committee for the Investigation of Atmospheric Pollution (1912)

Chairman:
Sir William Napier Shaw (*Director Meteorological Office*)

Hon. Secretary:
Mr J.S. Owens (*Coal Smoke Abatement Society*)

Committee members:
Mr C.J.P Cave (*Past President of the Royal Meteorological Society*)
Mr J.G. Clark
Professor J.B. Cohen (*Leeds University*)
Dr H.A. Des Voeux (*Hon. Treasurer, Coal Smoke Abatement Society*)
Dr Hawkesley (*Assistant Medical Officer of Health, Liverpool*)
Mr J.B.C Kershaw (*Hamburg Smoke Abatement Society*)
Dr R. Lessing
Dr E.J. Russell (*Director of the Rothamsted Experimental Station Harpenden*)
Dr E. D. Simon (*Smoke Abatement League of Great Britain*)
Bailie W. Smith (*Convenor of the Air Purification Sub-committee of the Glasgow Corporation*)
Mr S.A. Vasey (*The Lancet*)
Mr F.J.W. Whipple (*Superintendent of the Instruments Division, Meteorological Office*)

culprit identification into a more generalised concern with the nature and quality of a given city's air. The CIAP sought to extend the forms of atmospheric observation that took hold in British cities in the nineteenth century in both space and time. Through the effective use of widely dispersed and scientifically approved instruments it was hoped that a picture of British air pollution could be built up that connected atmospheric observations over great distances (and between different cities), while providing a much more intensive temporal record of pollution than had previously been possible.

A further sense of what the CIAP envisaged its role to be can be discerned in the following passage from their first report,

> For this line of development, precedent in various stages of completeness may be found in the registration of births and deaths, and marriages, the registration of the elements of weather, and in many other forms of economic enquiry. It is, indeed, the established line of development for nearly all forms of inquiry connected with the science of demography.[5]

This passage has particular pertinence for this book. By paralleling systematic air pollution observation with the sciences of meteorology and demography, it is clear that the CIAP wanted to give air pollution monitoring a new analytical and political standing. At an analytical level, it is clear that the Committee wanted air pollution monitoring to attain the same level of scientific credibility as meteorology. Meteorology had established its scientific credentials in Britain during the eighteenth century through the formation of common systems of climatic taxonomy and the systematic recording of the weather (see Golinski, 2006; Jankovic, 2000; Naylor, 2006). To these ends, the atmospheric epistemologies of meteorology, and its associated instrumental infrastructure, made it a paradigmatic science for students of twentieth-century air pollution. At a political level, it is also clear that the Committee saw air pollution monitoring as something that could have the same utility to the State as the accumulated knowledge associated with demography. As Foucault outlines, it was the science of demography, with its tireless accumulation of social statistics that enabled the notion of population, and associated systems of biopolitics and governmentality, to first emerge (Foucault, 2007 [2004]). By positioning the nascent science of air pollution at the intersection of meteorology and demography the CIAP clearly desired to form a system of what we would now term *atmospheric governmentality*. The new system of air power envisaged by the CIAP sought to combine a scientific understanding of the atmosphere as a physical environmental system with a keen awareness of the role of air in the biopolitical constitution and support of society.

If the CIAP embodied a distinct shift in the rationality and associated scale of air pollution government with science, it was realised at the inception of the Committee that it could only fulfil its new governmental ambitions through the effective development and distribution of an apparatus of air monitoring equipment. A key goal of the Committee was thus to 'draw up details of a standard apparatus for the measurement of soot and dust and the standard methods for use'.[6] It is important to realise, however, that the CIAP did not simply facilitate the administrative distribution of pre-designed devices (such as the deposit gauge), but constituted a framework of experimentation and scientific exchange for the development and adaptation of new and existing atmospheric devices in order to meet emerging scientific and governmental demands. According to Golinski, '[A]ll scientific instruments have their origins in experiments. But to become an instrument, a piece of apparatus has to cease to be experimental. It has to acquire a stable physical form, suitable for transportation and reproduction [...]' (2006: 110). In the development of what was essentially a form of governmental experiment, the CIAP actively devised instruments that not only met the scientific requirements of accuracy and reproducible precision, but also the heightened demands of durability, mobility and mass production associated with territorial government. Many of the instruments developed by the Committee found their origins in early instrumental designs of meteorology[7] and air quality analysis. Within the CIAP, however, the existing devices of air measurement had to be re-forged to meet the demands of a territorially expansive governmental science.

Before such instrument-based developments could proceed it was necessary for the CIAP to have a clear sense of what the instruments were to measure. The Committee noted at a very early stage that it had not been given a clear definition by the Coal Smoke Abatement Society of precisely what the 'air pollution' it was supposed to measure actually was. The early meetings of the Committee were thus embroiled in detailed discussions of what constituted atmospheric pollution. While much has already been said in this volume about the various cultural, medical and political definitions of pollution, what is important to note, in relation to the CIAP at least, is the role of instruments in actively shaping the working definition of air pollution. The Committee's discussions of air pollution began through an analysis of what constituted atmospheric purity. The Committee defined atmospheric purity on the following terms,

> Normal air may be regarded as consisting of a mixture of various gases in the following proportions:– Oxygen (20.94); Nitrogen (78.09); Argon (0.94); Carbon Dioxide (0.03); Helium, Krypton, Neon etc. (traces). In addition to these gases; there exists a quantity of water vapour, varying according to certain well-known physical conditions.[8]

While such a definition may not in itself be surprising, what is intriguing is the concept of *normal air* that is invoked. Foucault discusses the use of the *normal* and *abnormal* within the constitution of disciplinary apparatus of government and science (see Foucault, 2007 [2004]).[9] In the context of the CIAP, however, the idea of normal air is not offered as the paradigm towards which pollution abatement should be working, but merely as a calibrated baseline for the measurement of atmospheric change.

In opposition to this notion of normal air, the Committee developed the following understanding of pollution,

> The term 'pollution' in its widest sense may be regarded as applicable to anything that disturbs the above-described constitution of the air, but in the present investigation a more limited view was taken. For instance, an excess of the watery constitution, represented by excessive rain or an abnormal quantity of carbonic acid, may be regarded as pollution; but these constituents need no further consideration here. The Committee's interpretation of the term 'pollution' relates to such matter, solid, liquid, or gaseous, as reaches the surface of the earth or falls upon the buildings, &c., either by its own gravity or with the assistance of falling rain.[10]

While acknowledging that anything deemed to alter the state of normal air could be deemed pollution, the CIAP adopted a much narrower definition of pollution. By focusing on matter that 'reaches the surface of the earth or falls upon the buildings, &c., either by its own gravity or with the assistance of falling rain,' the Committee was essentially acknowledging the limitations associated with existing instruments for measuring air pollution. There are two basic ways in which to measure air pollution. Air pollution can be measured in suspended form, as a proportion of the ambient atmosphere it is in; or pollution can be calculated once it has been deposited from the air. The connections that the CIAP make between these two methods and rationalities of government is instructive,

> The investigation of atmospheric impurity may be dealt with in two ways:– (a) we may measure the amount deposited from air in a given time and area. (b) We may measure the amount suspended in the air at any given time and place. These aspects of the question have each their own special interest and importance; for example, the deposited matter is that which chiefly affects our buildings and vegetation, while the suspended impurity is responsible for smoke, fogs, and obstruction of light, as well as for certain deleterious effects on the respiratory organs.[11]

According to members of the CIAP to choose between the measurements of suspended and deposited forms of pollution was to differentiate between an architectural/ecological set of concerns (deposited pollution), and an interest in the impact of air pollution on human health (suspended pollution).

The problem that the Committee faced was that while interested in both suspended and deposited forms of air pollution, the instruments necessary for the accurate measurement and recording of suspended contaminates had not been adequately developed. Caught between the differing governmental goals that could be achieved by measuring either suspended or deposited atmospheric pollution, the actions of the Committee were inevitably guided by the instrumental capacities of the time. In the early years of the CIAP it was consequently the tried and tested deposit gauge (as prescribed by *The Lancet Study*) that would be the instrument of choice for the measurement of atmospheric abnormalities. While the CIAP justified its choice of the deposit gauge on the basis that all suspended pollution is ultimately latent deposited pollution, this justification clearly had to ignore much vaporous and gaseous air pollution (that would not take the form of solid depositions). Tacit recognition of this form of deliberate governmental amnesia is evidenced in the significant amount of support that the CIAP continued to give to experimental work on suspended pollution measurement. This curb to the early ambitions of the CIAP serves to remind us that governmental practices are not something that are guided purely by abstract rationality; they are also structured by the material limitations associated with the instruments of science. While such messy realities can often be smoothed over within intellectual accounts of governmental history, or by expedient governmental discourses, they should not go unrecorded.

Networking Instruments in Space: On the First Instrument-Based Survey of British Air Pollution

The circulations of governmental science

Having decided to focus, initially at least, on deposited forms of air pollution, the CIAP set about creating a network of instrument-based pollution monitoring in the UK. Having arranged for the manufacture of a standardised deposit gauge, the CIAP formed a partnership with 18 local authorities (and the Meteorological Office) to deploy and collect data from these instruments. This network of local authorities was not simply composed of cities that suffered the worst effects of air pollution. The network instead appears to have been based upon the cooperation of urban reformers and/or scientists who were convinced of the importance of pollution surveillance and reform. These varied actors came from a range of backgrounds including the sanitary reform movement, meteorology, urban planning, medicine and combustion engineering. It was the willing participation of these sympathetic authorities and individuals, as much as the mobile utility of the standard deposit gauge, that enabled the governmental science of air pollution monitoring envisaged by the CIAP to travel throughout Britain. It was also this loose collection of scientists and reformers who would provide the intellectual skein in and through which the CIAP pursued further instrument-based experiments for the recording of atmospheric pollution.

The first official measurements of air pollution taken by CIAP-approved devices were made in February 1914. Details of the partner authorities who worked with the CIAP in 1914 are provided in Table 5.2. The table illustrates that of the 39 gauges that were deployed by the CIAP in 1914 64% were located in Birmingham, London, Manchester and Sheffield. Despite this highly limited, not to mention uneven, geography of atmospheric surveillance, significant efforts had to be made by the CIAP to ensure that the results gathered from their first survey were compatible. Consequently, in addition to providing standard gauges, the CIAP also supplied a *Circular of Instructions* and a standard report form for analysts to record their results upon. The Circular of Instructions contained important information relating to the set up and location of the gauges, and guidelines on the best methods of collecting and analysing samples. The standard report form provided a vehicle for the recording of pollution levels on a monthly basis that could be returned to the CIAP for compilation and statistical analysis (see Table 5.3). Taken together these forms and manuals of instruction constituted

Table 5.2   List of CIAP gauge locations, numbers and analysts (1914)

Location	Number of gauges	Analyst
Birmingham	3	J.F. Liverseege
Bolton	1	Harry Hurst
Exeter	1	F. Southerden
Kingston-upon-Hull	1	A.R. Tankard
Liverpool	1	W.H. Roberts
County of London	6	J.H. Coste
Meteorological Office, London	1	S.A. Vasey (of *The Lancet*)
City of London	1	F.L. Teed
Malvern	1	C.C. Duncan
Manchester	10	E. Knecht
Newcastle-upon-Tyne	1	J.T. Dunn
Oldham	1	J. Warrington (Correspondent)
Sheffield	4	W.P. Wynne
York	1	S.H. Davies
Coatbridge	1	Messrs. R.R. Tatlock and Thompson
Greenock	1	J.W. Biggart
Leith	1	A. Scott Dodd
Paisley	1	R.M. Clark
Stirling	1	J.R. Watson

Source: Committee for the Investigation of Atmospheric Pollution (1916): viii, M.Off.Arch. MO 249 256

important circulatory devices that enabled instruments of atmospheric surveillance to operate effectively while separated by great distances.

Time does not allow for a detailed recitation of the processes of chemical and physical analysis that enabled local stations to complete all of the sections in *Form B* on a monthly basis. It is, however, worth reflecting upon the section of the form that deals with the issue of *Factor 'F' for the Gauge*. Factor Fs were developed for all gauges utilised in the CIAP's inaugural survey because of imperfections in the manufacture of the deposit devices. As the CIAP noted,

> [i]t will be observed that under 'gauge number' is given 'Factor F for Gauge'. It was found that all gauge vessels differed slightly in superficial area. They were intended to be four square feet in area, but owing to slight alterations in shape resulting from the process of enamelling with vitreous enamel, which

Table 5.3 CIAP report form (1914)

Form B

No. of Report........

COMMITTEE FOR THE INVESTIGATION OF ATMOPSHERIC POLLUTION
REPORT OF OBSERVATIONS FOR MONTH ENDING......19...

CENTRE.............................. GAUGE NO......... FACTOR 'F' FOR GAUGE......

VOLUME OF WATER COLLECTED..litres= ... MILLILETRES OF RAINFALL
TOTAL SOLIDS DISSOLVED.......grammes (dried @ 105ºC)
TOTAL INSOLUBLE MATTER ......grammes (dried @ 105ºC)
TOTAL SOLIDS COLLECTED....grammes=...tons per sq. kilometre

COMPOSITION OF UNDISSOLVED MATTER:–

	Grammes	% of total solids	Tons km^2
Soluble in CS$_2$ (tarry matter)	...	...	...
Combustible matter insoluble in CS$_2$	...	...	...
Ash			
Total undissolved matter....			

COMPOSITION OF DISSOLVED MATTER

	Grammes	% of total solids	Tons km^2
Loss in ignition	...	...	...
Ash	...	...	...
Total dissolved matter.....			
Sulphate as SO$_4$	...	...	...
Chlorine as Cl	...	...	...
Ammonia as NH	...	...	...

Remarks:–

Signed.. Date...

Source: Committee for the Investigation of Atmospheric Pollution (1916): viii, M.Off.Arch.MO 249 256

had to be done after the gauge vessels were turned to size, and the consequent warping, it was necessary to measure the catchment area of each gauge separately.[12]

Although sheet iron and enamel paint had been used for *The Lancet*'s deposit gauge, it was found that these devices did not last long in polluted urban environments (Brimblecombe, 1987: 149). It was decided by the CIAP to improve the durability of the deposit gauges through the use of cast iron and vitreous enamelling. It is clear, however, that in coating the cast iron gauge with vitreous enamel a number of inconsistencies were produced in the size and shape of different gauges. These inconsistencies would, of course, in the long term have seen devices exposed to the same quantities of air pollution producing very different final measures of contamination. The Factor F calculation was thus designed to allow for the degree to which a specific gauge differed from the standard measurements of the deposit device. While the Factor F calibration allowed the CIAP to gain comparable readings from partner authorities, the need for such standardising calculations reminds us that the objective desire of good government cannot simply be transferred to the unthinking materiality of the instrument. Objects, such as deposit gauges, and the practices that occur at their surfaces, have an active, not simply instrumental, role within the associations that collectively constitute government. To claim an active role for instruments within the processes of government is not, as Latour reminds us, to suggest that objects determine the actions of government. Instead, recognising the action of instruments within the various amalgamations of government acknowledges the differences they make to how government can be conducted (Latour, 2006: 63–86). It is in this context that much work must be done both with and on instruments (including constantly re-measuring, re-equipping, and re-calibrating) in order to support the cognitive capacities of government and science.

As noted earlier, much has already been written on the links between geography, technology and science. The validity of any scientific practice, experiment or measurement has always been connected to the ability of the procedure to be replicated under standard conditions by other people in different places (see Shapin & Schaffer, 1985). In relation to governmental sciences, the issue of technological replication becomes even more pronounced. The sense of responsibility for entire populations and territories that is typical of modern governmental rationalities has placed great pressure on the instrumental capacities of State knowledge gathering. In his book *Political Machines*, Barry (2001) describes the geographical strategies that are deployed by governmental authorities in the construction of technological systems of environmental knowledge gathering. According to Barry, the utility of technological instruments to States stems not from the capacity of the tool itself, but from its ability to ensure compatibility between all of the instruments a government has to deploy throughout its territory. In order to construct an accurate picture of the various qualities and characteristics of its territory, Barry describes how governments construct *technological zones* within which the compatibility of instruments is carefully monitored and enforced (2001: 3). As Naylor (2006) eloquently points out, science (and I would add in particular governmental sciences) is not only dedicated to the command of truth production, but also to the control of space.

It is because of the problems associated with the distribution of instruments like the standard deposit gauge in space that the circulation of instruments requires the establishment of networks of trust and verification. For the CIAP's first national survey of air pollution its systems of trust were placed in the designated analysts – many of whom were well known to CIAP members – who could be carefully instructed and guided in the practices of pollution monitoring.[13] These systems of inter-personal trust and support were built up through the regular meetings convened by the CIAP, the meetings of scientists and analysts at major clean air exhibitions (see Chapter Four), and through regular local site visits conducted by prominent CIAP members such as John Switzer Owens.

Locating early atmospheric science

In addition to the significant amount of work that the CIAP had to undertake in order to enable the spread of the standard deposit gauge in geographical space, considerable attention also had to be given to the geographical positioning of the gauges in local places. The sites associated with scientific practice and knowledge production have become the subjects of significant intellectual reflection in recent years (see Withers, 2001). Classical scientific practices have long been synonymous with absolute and Euclidean conceptualisations of space. In this geometrically inspired vision, science is depicted as operating in a kind of limitless pure space: a space that is enrolled in the formation and ordering of knowledge, but does not present an obstacle to scientific endeavour. To this end, the idea of situating and locating scientific practices has been utilised as a post-modern strategy for exposing and decoding the obfuscating ideologies of classical scientific discourse (see Haraway, 1991). While the idea of locating science has, at times, used the notion of location as a metaphorical device in order to uncover the broad-ranging financial, institutional and political contexts within which science operates, location can also be used as a more literal analytic for the scientific condition. The idea of locating science suggests the importance of recognising the complex locational geographies that inform scientific endeavours that are not conducted in epistemologically cordoned-off areas like laboratories and museums (see Livingstone, 2005). It is in the nature of sciences that are

dedicated to governmental ends to confront the reality of geographical existence in ways that other scientific practices do not. But when instruments are
expected to operate in non-standard, variable environments there is pressure
to both improve the precision and reliability of the instruments being used
and to select, test and designate the sites where instruments are to be located
in such a way as to minimise geographical variability.

The geographical challenges that faced the CIAP in 1912 were related to
how to ensure that deposit gauges were located in the urban spaces where air
pollution was a problem, while ensuring objective precision and the need
for regular service and repair associated with the instruments. The point is
that in order to be governmentally effective deposit gauges had to placed in
areas where they could best measure pollution events as they affected an
urban population (namely at ground level). In order to be scientifically effective, however, the deposit gauges needed to be in locations where they would
not be tampered with or receive abnormal dust inputs (namely in secure, but
open spaces). The locational rationalities associated with government and
science are not, it would appear, always complementary. The physical nature
of the deposit gauge further complicated the locational logistics of the CIAP
survey. With a catchment area of four square feet and a height from the
ground of four feet, deposit gauges were not things that could be left incon-
spicuously around city streets. Because of the governmental considerations
and scientific expectations of the CIAP survey, and the physical form of the
deposit gauge, much effort went into to assessing the potential sites that city
landscapes offered for air pollution monitoring. This was in effect an assessment of the lived spaces within which science would have to be retrofitted,
not a Euclidean area of limitless choice and desire.

Figure 5.2 illustrates the locations that were chosen in Sheffield for the
CIAP's first pollution survey of 1914. In the case of Sheffield, the four locations that were selected to house the standard deposit gauges were: (i)
Attercliffe Burial Ground; (ii) Hillsborough Park; (iii) Meersbrook Park;
and (iv) Weston Park. As with many other cities involved in the CIAP survey,
the Sheffield authorities chose burial grounds and parks as the main sites to
house their instruments. Burial grounds and parks provided ample room for
large gauges and also ensured that analysts could gain easy access in order
to sample from and service the equipment. The other advantage associated
with such locations was that they enabled deposit gauges to be located at a
distance from point sources of pollution, and thus enabled them to record
ambient atmospheric qualities for the cities in question. Other locations
utilised in the CIAP's survey in other cities included schools, hospitals and
universities/technology colleges. Each of these locations offered open spaces
and were often conveniently located for the analysts who worked in the
affiliated universities, colleges and hospitals. In some instances (particularly
with stations located in London) the rooftops of buildings were used to set

SHEFFIELD—(1) Attercliffe Burial Ground, (2) Hillsborough Park,
Map XI.   (3) Meersbrook Park, and (4) Weston Park.

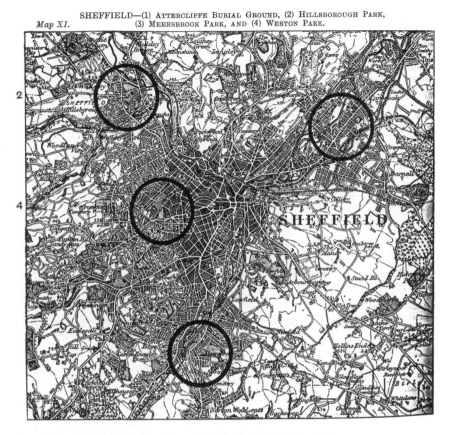

Figure 5.2   Standard deposit gauge locations in Sheffield for 1914 air pollution survey
Source: Committee for the Investigation of Atmospheric Pollution (1916): M.Off.Arch.MO 249 256

up monitoring stations. While providing relatively large open spaces in the very heart of the city, these locations were far from ideal, as they were obviously incapable of collecting pollution that was deposited before it reached the height of the elevated location. The physical properties of locations clearly matter in the production of governmental knowledge.

Collating and circulating the results of the 1914 air pollution survey

The complex instrumental practices, analytical procedures and locational decision making that surrounded the CIAP's first survey of air pollution in

Britain inevitably meant that the results were replete with errors and peculiarities.[14] Table 5.4 provides a detailed breakdown of the deposited samples that were collected by the 39 CIAP-approved devices that were deployed in 1914. This table of recorded measures is interesting for a number of reasons. First, it is important to note the use of the letters A, B, C and D within the records. These letters were calibrated to reflect measures that were above and below certain predefined levels. In relation to Tar, for example, A represented concentrations of less than 0.05 tons per km^2 per month; while the letter D was used to denote concentrations of over 0.25 tons per km^2 per month.[15] The value of these lettered codes stemmed from the fact that they provided the CIAP with a comparative framework within which they could immediately see how many As, Bs, Cs and Ds were recorded at different stations and thus discern broader geographical patterns. The second thing of note in Table 5.4 is the great diversities recorded for levels of rainfall between the different monitoring stations. If we compare the driest station (Leith with 38 mm of rainfall) with the wettest station (Greenock with 137 mm of rainfall), we have a range of rainfall levels spanning 101 mm. At one level such variations were not a major issue. The calculation of relative levels of atmospheric pollution in rainwater meant that the amount of rainfall did not affect the average records of contamination collated by the CIAP. The absolute levels of rainfall did, however, determine the temporal extent to which pollution could be monitored: with more rainfall normally facilitating a more detailed meteorological record of pollution events (allowing for variations in the intensity of rainfall of course). By using rainfall as a vehicle for the collection of air pollution the CIAP was essentially dependent upon natural levels in the fluctuation of precipitation to determine when pollution events were and weren't recorded.[16] It is in this context that the deposit gauge embodied an intriguing intersection between governmental surveillance, technological device and nature.

Technological Innovations, Suspended Pollution Monitoring and the Rise of Forensic Governmentalities

The CIAP and the Meteorological Office

Following the publication of the results of the first CIAP investigation of air pollution between 1914 and 1915, the procedures for atmospheric pollution monitoring associated with this survey were continued throughout the second part of 1915 and the entirety of 1916. The results of this wave of air measurements were compiled within the second report of the CIAP in 1917.[17] By 1917, however, concerns were being raised about the costs and sustainability of the CIAP study.[18] With the First World War raging on the

Table 5.4   Abridged 'monthly means of several elements of pollution' as recorded in the first survey of the CIAP

Station	Months	Rainfall (mm)	Tar	Ash	Sulphuric acid	Chlorine	Ammonia
Birmingham:							
Central	6	77	C	D	C	B	B
Aston	6	73	B	B	B	B	A
S.W.	6	84	A	A	B	B	A
Bolton	6	121	B	C	D	D	B
Exeter	6	88	A	B	B	B	A
Kingston-upon-Hull	6	64	B	B	B	B	B
Liverpool	6	75	D	C	C	C	D
London:							
Met Office	6	64	B	B	B	C	B
Embankment	6	66	C	C	C	C	B
Finsbury	6	65	C	B	B	B	B
Ravenscourt	6	64	B	B	B	B	B
Southwark	6	75	B	C	C	B	B
Wandsworth	6	56	A	B	B	B	A
Victoria	6	66	B	B	C	B	B
Golden Lane	6	73	B	B	C	B	B
Malvern	6	53	A	A	A	A	A
Manchester:							
Ancoats Hospital	3	78	B	D	D	C	B
Bowdon	3	69	A	A	B	B	B
Cheadle	5	66	B	B	B	B	B
Davyhulme	5	70	A	B	B	B	B
Fallowfield	5	70	B	B	B	B	B
Moss-side	3	75	B	B	C	C	B
Philips Park	3	71	B	C	C	C	B
Queen's Park	3	74	B	C	C	C	A
School of Technology	6	77	B	C	C	C	B
University	1	86	B	C	C	C	B
Whitefield	5	87	A	B	C	C	D
	6	63	B	C	B	B	C
Newcastle-upon-Tyne							
Oldham							
Sheffield:	6	94	D	D	C	C	D

Table 5.4   (cont'd)

Station	Months	Rainfall (mm)	Tar	Ash	Sulphuric acid	Chlorine	Ammonia
Attercliffe	6	66	C	C	B	D	B
Hillsborough Park	6	71	B	B	B	B	A
Meersbrook Park	6	75	B	B	B	C	B
Weston Park	6	73	B	B	B	B	A
York	6	62	B	B	B	B	D
Coatbridge	3	70	B	B	B	A	B
Greenock	5	137	B	B	B	C	C
Leith	5	38	B	B	B	B	C
Paisley	6	103	C	B	B	C	B
Stirling	3	120	D	B	B	B	B

Source: Committee for the Investigation of Atmospheric Pollution (1916): xxix, M.Off.Arch. MO 249 256

European continent, State finances and governmental resources were limited, while political interest in the work of CIAP was naturally muted. Despite being a largely voluntary endeavour it was estimated that the cost of the first two CIAP investigations was approximately £1000 per annum.[19] It was in relation to these costs that the CIAP submitted a successful grant application to the British government's Department of Scientific and Industrial Research (hereafter DSIR). As we will see, the DSIR would play a crucial role in the gradual governmentalisation of air pollution monitoring in the UK (see Chapter Six). But the most immediate consequence of this grant-in-aid was that the CIAP would have to be incorporated into a governmental body that would be responsible for administering the grant.

Given the fact that William Napier Shaw, the Director of the Meteorological Office at the time, was Chairman of the CIAP, it was decided that its most convenient home would be the Meteorological Office. Many British government officials were uneasy about this move of atmospheric pollution monitoring. It was felt that while meteorological insights could be helpful in understanding the kinetics of air pollution that the Meteorological Office lacked the core skills that were going to be necessary to accurately analyse the increasingly complex mixture of chemicals that constituted British air pollution.[20] It would, however, be the presence of Sir William Napier Shaw on the CIAP that proved crucial to its location within the Meteorological Office. Despite being best known for his research work in meteorology, Shaw had a long interest in issues of atmospheric pollution. These interests would, of course, culminate in the publication of his well-known tome, *The Smoke Problem of Great Cities* with John Switzer Owens in 1925 (Shaw & Owens, 1925). His passionate civic and governmental leadership on questions of British air pollution made him, rather than his Office, the logical guardian of atmospheric pollution inquiry in 1917. In coming under the auspices of the Meteorological Office the CIAP was required to change its name to the *Advisory Committee on Atmospheric Pollution* (hereafter ACAP). Although the movement of the CIAP into the Meteorological Office reflected the first incorporation of responsibility for systematic air pollution monitoring into the apparatus of national government in the UK, it was only the beginning of a long and acrimonious struggle over which government department should have responsibility for air pollution matters (see Chapter Six).

New technological developments at the ACAP and the colour coding of atmospheric science

With formal institutional support and allocated state funding, the renamed ACAP was able to dedicate more of its time to the refinement of existing technological apparatus, and the development of new instruments to aid atmospheric monitoring responsibilities. In relation to the refinement of the atmospheric instruments deployed by the CIAP and ACAP, it is important to recognise that the impetus for technological betterment was not something that merely radiated from a central committee to local sites of scientific practice. As described above, much has been written on the role of centres of coordination and calculation within the spatial extension of scientific practice. Within this work great attentiveness has been given to the flow of apparatus, instruction manuals and expert advice from the centre to the locality. It is evident, at least in the case of the CIAP/ACAP, that significant amounts of scientific insights and advice on technological adaptation actually flowed from local atmospheric monitoring sites to the centres of calculation and administration.

The annual reports of the CIAP and ACAP are full of feedback reports on the effectiveness of the deposit gauge within different local settings. The following is an example of one such report,

> Mr A.R. Tankard, of Hull, drew the attention of the Committee to this matter [errors and difficulties arising from algal growths]. The algal growth rendered filtration a tediously slow process, and to some extent was the source of error in the determination of undissolved matter and Ammonia [...] Several methods were suggested [for addressing the algal problem], the chief of which

were: the use of 1. Formaldehyde; 2. Mercuric chloride; 3. Metallic copper; 4. Copper sulphate. Mr Tankard kindly undertook to investigate the effect of formaldehyde. It was felt by him that the Ammonia determination might be interfered with (it was).[21]

It appears that the collection of standing water within the standard deposit gauge led to the production of large amounts of algal growths within the apparatus (particularly during the summer months). The presence of algae within the deposited samples made chemical analysis difficult and error prone. The chemical experiments carried out by A.R. Tankard on algae proved to be ineffective, but a method of decantation devised by Mr J.F. Liverseege (the air pollution analyst of Birmingham) did help with the filtration of deposited samples and was adopted by the ACAP as standard analytical practice.[22] The local experimental adaptation of monitoring techniques led to the regular reissuing of revised instructions for the analysis of rainwater deposits by the ACAP. While such processes of localised adaptation may seem innocuous, or even a natural part of the emergence of any experimental science, I believe that attentiveness to the technologies and spatial relations of governmental sciences necessitates a radical shift in ways we imagine governmental power. Governmental power over environmental objects like the atmosphere is not realised simply on the basis of the territorial extension of a standardised system of technological surveillance. Like all scientific practices, governmental science is always in a process of experimental becoming. While such processes of becoming are often hidden within the rigid certitude that surround the registers of governmental knowledge recording, attentiveness to the technological and spatial relations of government suggest the need for a less rationalist and more dilettante-based framework within which to interpret governmental science.

In addition to refining the standard deposit gauge, the new institutional standing of the ACAP, and its fresh funding resources, were also devoted to tackling the major lacuna within British air pollution monitoring: the measurement of suspended forms of air pollution. At a theoretical level the different ways in which it was possible to record suspended forms of air pollution had been known for some time. In a paper presented to the British Association in 1913, for example, John Switzer Owens outlined the different methods that could be used to measure atmospherically suspended pollution. The main methods outlined in 1913 by Owens were,

(1) A measured volume of air filtered through some medium and the deposit weighed; (2) Aitken's dust counter, by which the number of suspended particles can be counted; (3) A jet of air made to strike a glass plate and the opacity of the plate measured; (4) The opacity of a column of air measured in standard light conditions; (5) The visibility of fixed objects at fixed distances measured; (6) Impurities washed from air and weighed.[23]

John Switzer Owens' background was as a technical expert within the *Coal Smoke Abatement Society*. On the basis of his work for the Society he was made Honorary Secretary of the CIAP in 1912. Alongside Sir William Napier Shaw, Owens would prove to be one of the most influential scientists involved in the development of large-scale air pollution monitoring in Britain. With the support of the ACAP, Owens was at the forefront of developing and testing new procedures and instruments that would use the six available methods of suspended air pollution measurement.

Although the measurement of suspended forms of atmospheric pollution was relatively easy to achieve in individual experiments conducted by well-trained chemists, it did present particular problems to the establishment of larger-scale monitoring networks. As the amount of suspended pollution in a volumetric measure of air was relatively small, and the individual pollution particles finely divided, the early instruments of measurement were both expensive and difficult to operate. As was discussed in relation to the standard deposit gauge, high equipment costs and complex technological apparatus tend to militate against the formation of widespread isometric zones of atmospheric measurement. Working under the auspices of the ACAP, John Switzer Owens consequently used the following criteria when developing and assessing instruments for suspended air pollution measurement,

1. Simplicity of method so that it would be unnecessary to employ a skilled chemist to take observations. 2. Sufficient accuracy to give reliable, comparative, and, if possible, quantitative results. 3. Portability and cheapness of apparatus, as the value would be enhanced if widely used so as to give a large number of results at different places for comparison. 4. Speed of observation, which is a necessity, as the atmospheric conditions, especially in cities alter very rapidly. 5. Permanence of record for reference purposes.[24]

The particular requirements of governmental science meant that the instruments developed by Owens and his colleagues not only had to be sophisticated, reliable and accurate, but also portable, cheap and easy to use. It would be Owens himself who devised the first instrument to come close to meeting these exacting criteria.

During 1916 and 1917 Owens developed a relatively simple apparatus that was able to capture suspended air and provide a record of its suspended material.[25] The *Owens filter*, as it would come to be known, was based upon a standard filter paper, two glass bottles and a series of rubber pipes and bungs. By positioning one of the two bottles five feet below the other gravity was used to move water from one bottle to the other (see Figure 5.3).[26] This hydrological motion created a vacuum in the upper bottle that was then filled with air from the surrounding atmosphere as it was forced to pass through a filter paper. Through the use of gradation marks on the upper

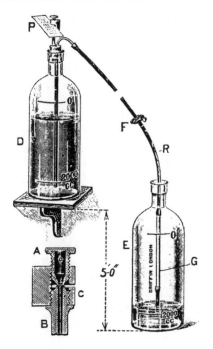

Figure 5.3   The Owens filter apparatus
Source: Committee for the Investigation of Atmospheric Pollution (1916): M.Off.Arch.MO 249 256

bottle it was possible for the analyst to ensure that precisely two litres of air had passed through the filter paper before the experiment was terminated. Such exactitude was obviously vital for the comparison of the discolouration of different filter papers across different devices.

The main difficulty with the technique of air measurement associated with the Owens filter was the issue of how results from different devices, and distant monitoring stations, could be compared. While the discolouration of filter papers was a convenient, and suitably simple, method for monitoring suspended pollution, the results that it produced could not be calibrated with the same exactitude as they had been for the deposited pollution collected in standard deposit gauges. The deposit gauge produced a precise, numerical record of air pollution (which could then be converted to an alphabetical system of reclassification). The Owens filter, on the other hand, produced a shade of colour not a number. Shades of colour are, obviously, far more difficult to convert into systems of comparable spatial data than figures, and as such it was necessary for Owens to devise a standard scale shade classification. Of course, in order to have a standard scale of shade to compare stained filter papers against, it was necessary to ensure that an accurate reproduction of colour could be produced in different sites.

In Chapter Three we observed the use of portable shaded charts to aid the classification of smoke density among air pollution inspectors. In relation to the work of the ACAP, and its desire to provide a less subjective, scientific basis for air pollution monitoring, a more robust colouration system had to be devised. Two problems confronted Owens when devising his system of shade and colour comparison. First, was the question of how to ensure that the colouration charts were produced with the necessary sharpness and accuracy to enable them to serve as calibration devices for air pollution surveys. Second, was the issue of what happens when colour travels though space. Little has been written on the impacts of geography on colour, but as soon as a colour moves from its point of production it is liable to be subject to the corrupting influences of fading and discolouration.[27] Distance is important here because it is essentially the barrier between a colour and comparison with its original. In order to address these problems, Owens devised an elaborate system of shade comparison that involved the layering of lampblack wash, which could be sent to different monitoring sites. The wash was designed so it could be applied to white paper every time an atmospheric reading was made: thus ensuring a level of colour consistency in the analysis. The numbered scale system used by Owen was 1, 2, 3, 4, 6, 8, 10, 12, 15 and 20, with the figures representing the number of painted layers that had to be applied to white paper to mirror the stain of the filter paper. But even with such an elaborate system of calibration, the Owens filter was undermined by two factors. First, it was predicated on the assumption that colours produced by pollution in different areas of Britain would be the same when passed through a filter. But variations in coal types and the forms of pollution produced in cement manufacturing districts served to challenge this basic assumption.[28] Second, and despite its attendant technological sophistication, by being reliant on colour comparison, the Owens filter inevitably brought the science of air pollution monitoring back into contact with the frailties of the observing eye.

Following the end of the First World War a series of new instruments and techniques for measuring suspended air pollution were developed in association with the ACAP. The Winkler tube and tintometer, for example, were both developed as a way of chemically differentiating between the gaseous and solid constituents of suspended air pollution (a facility that was not part of the Owens filter) (see Figure 5.4).[29] By comparing air drawn directly from the atmosphere into a solution of methyl-orange with air that had been previously been passed through the Owens filter (in order to remove its solid content) the Winkler tube and tintometer made it possible to determine the relative constitution of non-solid, suspended air pollution.[30] Through

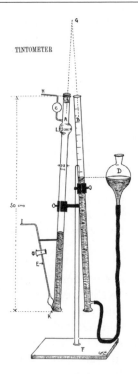

Figure 5.4   The tintometer
Source: Committee for the Investigation of Atmospheric Pollution (1916): M.Off.Arch.MO 249 256

the use of complex titration apparatus, the tintometer was not only able to provide a relative measure of solid and gaseous suspended pollution in the air, but also a very precise reading of the acid content of suspended gaseous pollutants. What marked out the Winkler tube and tintometer from other devices used in the measurement of suspended atmospheric pollution was that they deployed chemical procedures and tests. The mixing of politics and chemistry within the apparatus of air pollution monitoring reminds us that the governmental production of atmospheric knowledge was not only the product of material technologies, but also depended on the effective marshalling of established chemical tests and procedures. Interestingly, the mixing of government and chemistry embodied in the Winkler tube and tintometer did not entirely overcome the problems of subjective sight that were a feature of the Owens filter. In order to determine accurate measurements for the concentration of suspended pollution in the tintometer, for example, it was necessary to be able to detect very accurately the point at which a methyl-orange solution turned from yellow to red. The chemists working with these apparatus complained that after long periods of working with the devices it became difficult for the eye to detect the subtle changes in colour associated with the process.[31] This *fatigue of the eye* was another threshold displacement between objectivity and subjectivity in the evolving relationship between state, science and the atmosphere in Britain during the early twentieth century.[32]

Counting pollution and forensic governmentalities

The inability of the Owens filter method to produce numerical data, and the problems of colour identification associated with other forms of chemical air analysis developed within the ACAP, limited their respective governmental utility. These problems were, however, resolved by John Switzer Owens in the development of his jet dust counter (see Figure 5.5). The jet dust counter was a model of the type of sophisticated simplicity that the ACAP were looking for in its suspended air pollution monitoring technologies. Comprising a small box with what appeared, to all intents and purposes, a coin slot on its front, the jet dust counter worked by drawing a standard amount of air through the slot at regular intervals. The air sample then passed through a damping chamber before being collected on a piece of glass. The cover glass could be removed by simply inverting the whole device and its surface analysed by use of microscopes of varying strengths.[33] Under microscopic analysis pollution particles collected were counted over a cross-section of standard size so that a comparable numerical value for the density of solid pollution could be ascertained. In addition to providing a measure of pollution, the jet dust counter also made it possible to record the size, shape and distribution of particles according to different pollution events. In the context of these dual benefits, dust counts facilitated the emergence of new regime of micro-forensic govermentality in Britain.

In order to appreciate the nature of the micro-forensic forms of governmental knowledge that were ushered in by the Owens jet dust apparatus it is helpful to reflect on the following ACAP record of different dust count events,

During the dense fog of January 22, 1922, a record of 50 c.c. taken gave 21,750 particles per cubic centimetre. A large proportion of these particles were 1.7 microns in diameter, while the average diameter was 0.85 microns. The number of large-sized particles found during this fog was quite unusual, as such particles are usually very few. During the fog of October 26, 1921, in London, the number of particles per cubic centimetre was 20,800, the average size of the particles was 0.85 microns and the maximum was 1.7 microns.[34]

The ability to gain such precise and detailed data on the nature of suspended air pollution facilitated a new forensic mentality within the

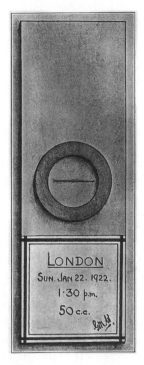

Figure 5.5   The Owens jet dust counter
Source: Committee for the Investigation of Atmospheric Pollution: M.Off.Arch.MO 249 256

government of atmospheric pollution in the UK. This forensic mentality was in part based upon the new-found capacity to compare the severity of pollution events in different places on the basis of the relative intensity and size of particles suspended in the air. Suddenly, pollution events that could have appeared remarkably similar when recorded in aggregate forms in deposit gauges could be accurately differentiated, assessed and potentially addressed. To these ends, the Owens jet dust counter essentially enabled the ACAP to look into the pollution event as it was happening: to monitor pollution in its air-borne form.

In addition to facilitating the more accurate classification of air pollution events, the Owens jet dust counter also facilitated a forensic mentality within the identification of the sources of air pollution. Below, for example, is the following account of the use of the jet dust counter in the English county of Norfolk in 1921,

> In comparatively pure air the volume drawn through the jet may have to be increased, for example, in samples taken during last August on the coast of Norfolk, it was found necessary sometimes to draw 1,000 c.c. to obtain a

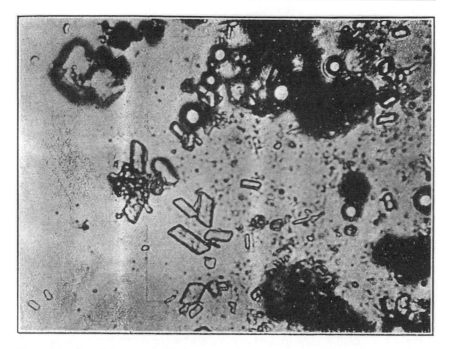

Figure 5.6   Record of country air of Surrey taken by jet dust counter showing crystals from a dried-up stream bed
Source: Committee for the Investigation of Atmospheric Pollution (1916): M.Off.Arch.MO 249 256

suitable record, and during a slight haze in dry sunny weather, about 100 to 200 particles per cubic centimetre were found, the size ranging from about 0.3 to 1.7 micron. These were found during a North East Wind and must have been carried across the North Sea from the Continent of Europe.[35]

While the prevailing wind provided the key clue to the source of Norfolk's pollution in 1921, the high resolution vision of pollution provided by the jet dust counter meant that even relatively imperceptible forms of air pollution could be matched with existing pollution signature types, and attributed to likely local (or more distant) sources. The level of forensic scrutiny facilitated by the jet dust counter was further enhanced when it was realised that once collected, the water that was used to capture dust particles could be dried to reveal crystals produced by other chemical pollutants. In order to assist in the reading and classification of crystal forms a standard series of slides were produced for ammonium chloride, potassium chloride, ammonium nitrate and sodium sulphite (among many others), which could be directly compared under a microscope with crystals collected from the jet dust counter (see Figure 5.6).

The high level of microscopic precision associated with the jet dust counter meant that it not only enabled the emergence of a forensic form of
governmentality, but also facilitated the study of the biopolitical aspects of
air pollution in new ways. As we have already stated, the desire of the ACAP,
and erstwhile CIAP, to study suspended atmospheric pollution was premised on the fact that air-borne pollution had most direct relevance to
questions of human health. Despite providing insights into the nature of
suspended pollution, however, the Owens filter, Winkler tube and tintometer
were not sensitive enough to analyse the nature of the interaction between
air pollution and the human body. Once developed, however, the Owens jet
dust counter provided a technological apparatus that was responsive enough
to monitor the transferral of pollution from the atmosphere and into the
lungs of the human body. To this end, the jet dust counter was not only used
to assess the nature of public air, but was also deployed as a quasi-medical
instrument of bodily assessment. In the early 1920s the ACAP commenced
a series of controlled experiments that used the Owens jet dust counter to
assess the difference between ambient air pollution and the air that was
leaving human lungs. This research sought to build on the groundbreaking
work of John Tyndall, who, in the second half of the nineteenth century,
conducted some of the earliest research on the nature of expired air.[36]
Although Tyndall's experiments on breathing are often synonymous with
the conclusion that expired air is dust free, his results only indicted that air
drawn from the depths of lungs was pure. Tyndall was not, however, able to
provide an accurate assessment of the amount of contaminates that stay in
the human body following an intake of polluted air. Through careful experimentation with the jet dust counter, Owens was able to ascertain that on
average only 70% of the total dust particles that entered human lungs left
the body when the air sample was exhaled.[37] Owens presented his findings
at a meeting of the Medical Society in London on 12 December 1921.[38]

Owens' medical research was important on two fronts: (i) it ascertained
accurately the amount of particulate pollution that could be expected to
remain in the lungs and cause related respiratory illnesses; (ii) it conclu-
sively demonstrated that breathing through the nose made no difference to
the amount of dust that entered the human lung.[39] This second analytical
insight was of particular historical importance. After it was realised in the
nineteenth century that smoke pollution was not a helpful disinfectant it
was still argued that many of its worst health effects could be prevented by
the use of the nose as a kind of natural filter for the air.[40] Studies conducted
by the Owens jet dust counter finally put this olfactory theory to rest, and
confirmed medical concerns that smoke was not just a general public
nuisance, but a real and present threat to public health.

This chapter has sought to emphasise that changing governmental attitudes
towards atmospheric pollution (like those associated with a forensic rationality)

cannot be understood in isolation from the technological developments and devices that run through the history of air pollution government. While attempting to uncover the agency of instruments in the history of atmospheric government, however, I am not trying to argue that technological devices change history. What is rather at stake here is whether, when we trace histories of governmental reason, we choose to depict technological things as *intermediaries* or as *mediators* (Latour, 2006: 39–42). When understood as intermediaries – or that 'which transports meaning or force without transformation' (ibid.: 39) – knowing the governmental reasons and forces behind the original development of the jet dust counter is enough to understand its eventual governmental utility. When analysed as a mediator, however, Latour reminds us that the ultimate role of devices like the jet dust counter cannot easily be determined from the inputs that shaped their creation. It is in this context, that the jet dust counter, as with so many of the other devices developed and deployed by the CIAP and ACAP must be interpreted as mediators of, not intermediaries for, government. Developed initially as part of the broader scientific struggles to accurately measure suspended air pollution from place to place, the forensic potential of the jet dust counter clearly changed what it become possible to govern in British atmospheric relations. The experiments conducted using the jet dust counter also changed prevailing governmental attitudes concerning the likely health implications of air pollution and ushered in a new age of State action with regard to atmospheric relations.

Coda: From Governing Things to Governing Through Things

In this chapter I have argued that histories of atmospheric (or indeed any form of) government should not only seek to uncover the temporal genealogies of governmental rationality, but also the role of technological things and processes in affecting governmental history. Consequently, through an analysis of the Committee for the Investigation of Atmospheric Pollution (and the later Advisory Committee on Atmospheric Pollution), and the associated deployment of the standard deposit gauge, Owens filter, Winkler tube, tintometer and jet dust counter, this chapter has explored the role of instruments within the history of British atmospheric government. Attentiveness to the significance of instruments is not uncommon within historical accounts of science, and is a regular feature of science and technological study. Far less has, however, been written on the role of technological devices within histories of government. In this coda I want to consider two questions: (i) why have studies of governmental history paid so little attention to the role of instruments within the evolution of governmental techniques; and (ii) does it really make any difference if we chose to tell stories of technology within our studies of government history?

The answer to the first question in many ways requires us to first abandon the premise of the question. The role of technology has appeared within accounts of contemporary and historical government: authors have consistently emphasised the role of cartographic equipment, navigational devices and medical apparatus within the consolidation of governmental power and influence (see Edney, 1999; Scott, 1998). The work of Foucault on biopolitics and governmentality is strewn with reference to various manifestations of governmental instruments (including accounts of technologies of surveillance, survey and sanitation). The point is that within existing accounts of governmental history the role of technology tends to be reduced to that of the tools of existing governmental desires and rationalities. This point leads directly into the second question. In asking whether governmental histories would really be any different if told from a non-technological point of view, I am really asking the question: why is it necessary to ascribe agency to technological things? Given the contemporary desire within the social sciences to rapidly (and sometimes unthinkingly) expand our sense of what has agency, this question has a particular import. According to Latour, a commitment to recognising the mediations of things should not be premised on some form of ideologically predisposed monism, within which we are ethically impelled to recognise the equal role, and thus worth, of all things. Rather this dedication to uncover the role of things within politics, science, history, economics and cultural formations of different kinds is a methodological commitment. As a methodological commitment, the study of things supports a form of analysis that is resistant to looking for answers to questions of historical change within explanatory categories like nature, the social, or the State and science. It is in this methodological context that this chapter's focus on the instruments associated with British atmospheric governmentality is so important. By considering instruments as mediators of governmental desire, it has been possible to see how the production of knowledge, and associated strategies of atmospheric government, have not simply flowed through a hegemonic rationality of and for socio-environmental governance. What it is possible to know and govern within the British atmosphere has often been the outcome of the unforeseen utility of new and existing instruments in providing novel insights into the nature and extent of air pollution.

In a thought-provoking synopsis of the work on the history and sociology of science, Steven Shapin reflects upon the question of whether any related study of science has ever been able to reveal how contingent factors (such as the design, replication and durability of instruments) actually changed the nature of the reality that scientists ultimately exposed (1995). At one level, and as Shapin himself acknowledges, an appreciation of the material contexts and locations within which scientific knowledge is produced may not prove how a different version of reality would have been fabricated under different conditions. But in the context of the types of territorially expansive governmental sciences explored in the chapter, it is clear that the durability, cost of replication, mechanics of calibration and locational opportunities associated with air pollution monitoring technologies did have an impact on the geographical extent and placing of scientific knowledge production. In this sense it is clear that sensitivity to the nature, and agency, of instrumentation enables us to better understand why certain forms of atmospheric knowledge were produced while others go unrecorded and unremembered.

Chapter Six

A National Census of the Air: Spatial Science, Calculation and the Geo-Coding of the Atmosphere

The previous chapter considered the role of instruments in the spread and consolidation of the nascent governmental science of air pollution monitoring in Britain. While focusing on the various instruments that were developed by the CIAP and ACAP, Chapter Five considered a series of spatial issues – particularly the location and circulation of devices – that were crucial to the 'tooling up' of British air pollution science. Moving beyond geographical questions of location and circulation, this chapter explores how the British atmosphere gradually became an object of spatial calculation. Analysis argues that from the late 1920s onwards the British atmosphere became increasingly subject to a twin process of *geo-coding*. At one level this geo-coding involved the growth of an ever more diverse and extensive geo-historical record of air pollution. As more long-term registers of air pollution rates were kept for more places, it became possible for government scientists to compare pollution records between different locations, and develop new governmental understandings of the reasons for the spatial variations observed in pollution rates. At a second level, this geo-coding process involved the use of standard spatial units (such as the atmospheric region and the British standard measure of air pollution) to delimit and calibrate air pollution measurement. Both of these geo-coding processes were based upon distinctive forms of spatial science and rationality and were central to the calculation of atmospheric affairs and the development of associated techniques of air government.

A central tenet of this chapter is the assertion that space has been both a barrier to the calculation of the atmosphere, but also a valuable framework for air measurement. A series of contemporary writers on the State expose the important connections that exist between calculation and the spaces of government (see Scott, 1998; Hannah, 2000; Elden, 2007). At one level, the relationship between government, calculation and space can be discerned in the ways in which State territories have literally been produced by the surveys, cartographic projects and land registers that have inexorably calculated different States' spatial being (Pickles, 2004;Whitehead *et al.*, 2006: 86–116). While much has been written on the calculative view of space promoted within modern State systems,[1] far less has been said about the use of calculable epistemologies and rationalities of space as the basis for constructing frameworks within which to conduct projects of government. When constructed as something that is stable, bounded and finite this chapter argues that space has not only been the object of governmental calculation, but has increasingly been the subject of different calculative campaigns.

This chapter begins by returning to the political struggles that surrounded the funding and operation of the ACAP during the 1920s (see previous chapter). With new specialist support and institutional funding, analysis considers how the ACAP aspired to generate a nationalised science of air pollution monitoring in Britain (particularly in relation to the creation of a British standard measure of air pollution), despite the spatial difficulties associated with such a scientific project. The following section moves on to consider the impact of the Second World War on air pollution science in Britain. This section claims that while placing a significant strain on the monitoring network already established in Britain, emerging forms of military interest in air pollution generated a new vertical territorial perspective on the atmosphere. Analysis then considers the impact of what is seen by many as the most significant event in the history of British air pollution: the London fog disaster of 1952. Following the London fog disaster and the subsequent Clean Air Act of 1956, the next section considers the formation and implementation of the first national survey of air pollution ever conducted in Britain. The National Air Pollution Survey ran from 1961 to 1971 and was coordinated by the Warren Spring Laboratory. While this chapter is interested in how the National Air Pollution Survey generated a new synchronised space of national air pollution monitoring, analysis focuses upon how space was used by the survey to provide a new framework for atmospheric calculation. Ultimately analysis uncovers how the science and government of atmospheric pollution in mid-twentieth-century Britain met within a common scientific deployment of space.

The Spatial Expansion of Air Pollution Monitoring in Britain

Pollution science in question and the re-institutionalisation of the ACAP

The previous chapter described how the CIAP, followed by the ACAP, developed a range of instruments and standardised procedures that facilitated the spatial expansion of air pollution monitoring in Britain. Despite the significant technological advances made by John Switzer Owens, the ability of the ACAP to expand monitoring activities across Britain came under threat during the 1920s. The problems began in 1920 when the Meteorological Office became amalgamated into the Air Ministry.[2] One of the consequences of this amalgamation process was that the grant aid that the ACAP received from the Department of Scientific and Industrial Research stopped, and responsibility for funding the Advisory Committee was transferred to the Air Ministry. While the Air Ministry was able to take on the financial costs associated with the work of the ACAP, difficulties started to emerge in 1924 when, with the spatial scope of their work and operations expanding all of the time, John Switzer Owens asked for additional funding to support the work of an air pollution analysis laboratory in London.[3] With direct oversight on the ACAP being transferred from Sir William Napier Shaw (of the Meteorological Office) to Sir William Nicholson (Chairman of the Air Ministry), Owens found a far less receptive audience for his funding request. Sir William Nicholson was wary of offering additional financial support to the ACAP for two main reasons: (i) he did not approve of spending government money on 'speculative scientific research' that did not have a clear practical application; and (ii) he did not feel that the work of the ACAP addressed issues of meteorology enough to be eligible for funding as part of the Meteorological Office.[4]

In order to resolve the tension between Nicholson and the ACAP the *Committee of Enquiry into the Future of the Advisory Committee on Atmospheric Pollution* was established. Additionally, an Interdepartmental Conference of the British government was convened on 25 April 1925 to discuss the ACAP and to supplement the work of the Committee. This Committee and Interdepartmental Conference raised two crucial issues regarding the ACAP. The first related to the fact that the work of the ACAP was clearly not directly relevant to the broader responsibilities of the Meteorological Office or Air Ministry and thus needed to be relocated within the institutional apparatus of the State.[5] The second, and more problematic issue, concerned the value and nature of the science conducted by the ACAP. In relation to the second set of concerns, questions were raised among leading civil servants and government officers about whether the work of the ACAP was something that the government should support at all.[6] At one level this question derived from the uncertain practical benefits of the ACAP's work that Sir William Nicholson had highlighted. In a meeting convened immediately before the Interdepartmental Conference, for example, it is recorded that Mr Gibbon of the Ministry of Health,

[e]xpressed himself in sympathy with scientific research, but thought the financial assistance of the state should be reserved for more practical matters than the work in question. If he were offered £2,000 per annum he could find

better use for it (I afterwards learnt that he would apply it to research on activated sludge treatment of sewage).[7]

While the reflections of Mr Gibbons raise interesting questions about the types of science that a State should or should not support, his basic sentiment was that State science should be dedicated primarily to subsidising problem-solving technologies, not knowledge-gathering apparatus; a sphere that Gibbons perhaps associated more with the types of sciences that were practised without direct State influence and support.

Beyond such ideological assertions about the nature of State science, the primary critique of the atmospheric sciences being practised by the ACAP was that they were not scientific enough. It emerged during discussions about whether the ACAP should be transferred into the Department for Scientific and Industrial Research (hereafter DSIR) that many in key government departments were highly sceptical of the veracity of the scientific knowledge being produced by John Switzer Owens and his colleagues.[8] These concerns were partly derived from Owens' involvement with the Coal Smoke Abatement Society. Given that the Coal Smoke Abatement Society was essentially a single-issue lobby organisation, there was concern within government circles that Owens was simply using the ACAP as a vehicle to give scientific credence to a broader smoke abatement movement.[9] In addition to questions of ideological bias, however, concerns were also raised about the reliability of the scientific procedures and practices that had been developed by Owens under the auspices of the CIAP and ACAP. In a letter written by L.S. Lloyd of the DSIR on 4 March 1926, these concerns were addressed bluntly,

> I am left with an uneasy feeling as to the position in which we will find ourselves scientifically if we take over Dr Owens and his work. The impression left on my mind after yesterday's meeting was that the Council did not really think much of his work; you will remember the illusion to amateur enthusiasts. I gathered quite recently that Dr Simpson [Director of the Meteorological Office] has some lack of confidence in the reliability of Dr Owens' methods and competence to interpret his own results. You may be able to have words with him about this.[10]

According to Moseley (1980), since its creation during the First World War the DSIR (initially the Committee for Scientific and Industrial Research) had placed great emphasis on the importance of supporting non-partisan, independent science of the highest standing.[11] It appears that the ACAP was seen by many in the DSIR at the time to have dubious standards of both independence and scientific method.

Despite these significant reservations, it was widely felt that if the ACAP was to find a home in government at all it should be in the DSIR. The

DSIR was formed to aid commercially prohibitive research that could have long-term benefits for the British economy, and to oversee the operations of the National Physical Laboratory (see Moseley, 1980; Rose & Rose, 1971: 40–6). This designated role meant that the DSIR had the scientific expertise and experience of developing, testing and distributing scientific equipment that the ACAP appeared to require. The first request for the DSIR to take responsibility for the ACAP came from the Air Ministry in 1925.[12] This initial request was turned down by the DSIR on the grounds that the research carried out by the ACAP did not have the potential to develop the practical technologies that could help British industry address pollution issues (a priority that was central to its institutional mission).[13] However, the pressure for the DSIR to assume responsibility for the ACAP continued unabated. At one level pressure was placed on the DSIR in a very personal form, with key dignitaries of the ACAP, like William Napier Shaw, attempting to persuade personnel at the Department of the Committee's value. In one such attempt Shaw sent a complimentary copy of his and John Switzer Owens' book, *The Smoke Problem of the Great Cities*, to L.S. Lloyd, Director of the DSIR, as an example of the exemplary work and techniques of the ACAP. Lloyd's response to this bequest perhaps underscores the DSIR's suspicion of the ACAP,

> It was most kind of you to send me your book on the Smoke Problem of Great Cities which I shall read with great interest. I feel that it is rather heaping coals of fire on my head since I am afraid the proposal to transfer responsibility for research into atmospheric pollution has been made impossible by the Economy campaign.[14]

At another level, the planned 1926 Public Health (Smoke Abatement) Act placed a series of legislative pressures on the DSIR to assume responsibility for the ACAP.[15] The proposed Act had two implications for the scientific study and monitoring of air pollution in Britain. First, it extended the coverage of air pollution legislation from black smoke to all smoke nuisances issuing from industrial premises.[16] This legislative shift generated the need for the development of accurate and easy-to-use devices that could help assist the ever-complex work of local smoke inspectors. Second, the proposed Act also made provision to support local authority based research into techniques for pollution measurement and monitoring.[17] In light of these legislative changes it was felt by many involved in atmospheric government that the DSIR, in partnership with the ACAP, could effectively support the 1926 Act.

The lobbying activities and legislative pressures that accrued around the DSIR during 1926 led to the Department eventually assuming responsibility for the management of the ACAP.[18] The DSIR did, however, establish certain preconditions that were attached to its leadership of the Committee.[19] The first precondition was that the ACAP's dual role in air pollution monitoring and the development of new technologies to assist with such surveillance activities become more clearly demarked. This led to the subdivision of the ACAP into the *Atmospheric Pollution Research Committee* and the *Standing Conference of Cooperating Bodies*.[20] The Atmospheric Pollution Research Committee would assume responsibility for the development of new technologies associated with matters of air pollution. Significantly the remit of this new committee was not exclusively focused on advancements in monitoring technologies (as had been the case with the ACAP), but was instead defined more broadly as, '[T]o carry out special researches directed towards the solution of the scientific problems of atmospheric pollution'.[21] The redefinition of the technological role of the ACAP brought it in line with existing research being carried out by the DSIR's Fuel Research Committee.[22] The Standing Conference of Cooperating Bodies (hereafter SCCB) was designed to support and extend the network of air pollution monitoring established by the CIAP/ACAP. Given the tight financial restraints that surrounded the DSIR, the Department envisaged the SCCB acting in, '[a] consultative and advisory capacity to bodies cooperating in the research'.[23] In this context, although the SCCB would offer advice and monitor the procedures and practices of local air pollution monitoring stations, the costs of air pollution monitoring and equipment supply would fall on cooperating bodies. Despite concerns over his previous research record, John Switzer Owens was appointed *Superintendent of Inspections* for the SCCB, and among his many duties he was responsible for regularly testing the equipment used to monitor air pollution in different localities. Also in keeping with the ACAP's hierarchy, William Napier Shaw became the Chairman of the SCCB.[24]

Expanding atmospheric science and the operation of the SCCB

The SCCB first met on 23 April 1928. This meeting brought together Sir William Napier Shaw, John Switzer Owens, members of the DSIR and representatives from participating local authorities.[25] The primary aim of this meeting was to agree on the methods of air pollution measurement that would be deployed by Conference members and to consider the most effective ways of extending atmospheric surveillance throughout Britain. A circular produced at the time by the DSIR and entitled *A Note on the Investigation of Atmospheric Pollution* reveals the underlying desires and concerns of those leading the SCCB,

The object of the investigation is to obtain exact information about the nature and extent of atmospheric pollution [...] Unless a considerable number of authorities take part in the observations using the standard methods and appliances, the data will be imperfect. The information obtained must be definitive, and sufficiently complete, if it is to be possible to fix standards for clean air (DSIR, 1928: 1).

It was decided by the SCCB that harmonisation in atmospheric surveillance (and the avoidance of imperfect data production) would be most effectively pursued by using the instruments that had been tried and tested by the CIAP and ACAP. The territorial expansion of atmospheric monitoring was thus to be facilitated by building on the existing network of local authority partners and by canvassing other key spatial constituencies to become members of the SCCB. The general costs of maintaining a network of air monitoring stations was a deterrent to many local authorities participating in SCCB activities. Beyond general costs, however, the fact that the SCCB was funded by voluntary contributions from members meant that even local authorities that did monitor air pollution had good reason for not joining the SCCB.[26] Notwithstanding these factors, it is clear that during the 1920s the SCCB significantly extended its air surveillance capacity. In a written response to a question raised in the House of Commons, Neville Chamberlain (then Minister for Health) stated that there were 40 local authorities cooperating with air pollution monitoring activities by 1929.[27]

In 1930 a map of participating local authorities was produced by the SCCB (see Figure 6.1). This map was displayed around the country at clean air exhibitions and government conferences. This cartographic exercise reveals a number of interesting things. First it shows not only the location of the local authorities (and other bodies) that were taking routine readings of air pollution levels, but also the number of devices that were being deployed in the area (even differentiating between automatic and non-automatic filters). Second, the map also distinguishes between three types of local authority: (i) those that made atmospheric observations; (ii) those that made atmospheric observations and contributed financially to the running of the SCCB; and (iii) those that made financial contributions, but did not keep records of atmospheric pollution. Beyond these interesting cartographic details, however, what is perhaps most important about this map are the governmental intentions that clearly informed its production.[28] At one level it was clearly designed to serve a disciplinary purpose. By differentiating between those areas and organisations that helped to fund the work of the SCCB, and those that did not, the map reflects a kind of moral geography of responsibility/irresponsibility that it hoped would stimulate more wide-ranging support for the SCCB. Most importantly, however, the map reveals the expansionist spatial desires of the SCCB. By mapping the sites

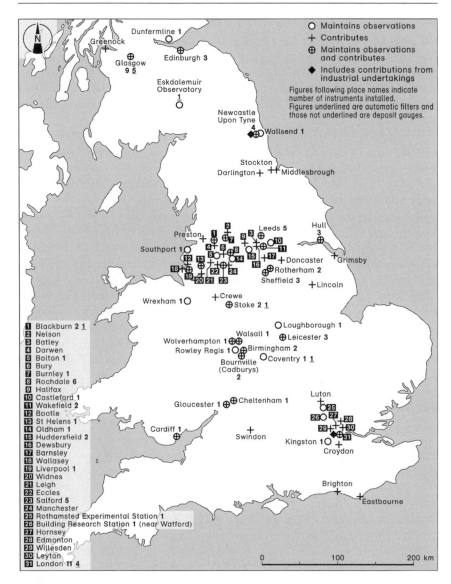

Figure 6.1   Map showing location of ACAP members in 1930
Source: TNA.DSIR14/2

and locations of atmospheric surveillance the SCCB was able to reveal the highly uneven spatial development of air monitoring activities in Britain during the 1920s and 1930s. The map clearly reveals the concentration of atmospheric surveillance activities around the large conurbations of London, Birmingham, the English northwest and Glasgow. While illustrating the concentration of monitoring efforts around the main sources of pollution, the map reveals no corresponding apparatus of surveillance in large cities like Bristol, Nottingham, Exeter, Reading or Swansea. In addition to the absence of such larger settlements, the map also reveals large spatial gaps in monitoring coverage, with no stations in East Anglia, Kent, and large areas encompassing Oxfordshire, Cambridgeshire, Worcestershire, Herefordshire and Northamptonshire.

The glaring spatial lacuna within British atmospheric surveillance, which was identified within the SCCB map of 1930, meant that not only was there a lack of standardised knowledge concerning levels of air pollution from key polluting sites (Bristol, Nottingham etc.), but that it was also difficult to understand the trajectories and dispersion patterns associated with air pollution (the use of non-urban air monitoring stations can provide useful insights into the prevailing directions in which pollution is dispersed once it leaves its point of production). This map was thus not so much a celebration of the developments that had been made in British atmospheric surveillance since 1912, but a statement of governmental intent concerning what still needed to be done.[29] Following the production of this map, the 1930s became a period of significant expansion in the monitoring of air pollution in Britain. In 1930 there were 92 SCCB-approved monitoring sites operating in Britain.[30] By the outbreak of the Second World War the DSIR estimated that there were 225 standardised governmental devices dedicated to the monitoring of air pollution in Britain.[31] Even with the obvious disruptions generated by the war effort, by 1949 this figure had risen to 538 gauges, filters and assorted instruments.[32]

The results of the expanding spatial network of monitoring devices supported by the SCCB were published in monthly *Atmospheric Pollution Bulletins*, which were compiled by the Fuel Research Station of the DSIR. As discussed above, analysis of various forms of governmental survey stress that the territorial expansion of surveillance facilitates the development of a more complete picture of governmental responsibilities and potentials. While it is clear that the work of the SCCB during the 1930s and 1940s enabled the production of a more spatially complete picture of air pollution in Britain, it also enabled other governmentally useful calculations of the air. As the temporal records of air pollution at different sites accumulated it became possible for the SCCB and different participating bodies to aggregate and compare different spatial data sets of air pollution. By being able to compare changing levels of air pollution between different places over time, it became possible to consider why atmospheric pollution increases in one place while remaining stable or decreasing in another. The search for answers to these questions has important implications for atmospheric government, revealing as it often does the most effective government strategies to deploy when combating pollution events.

The geographical categorisation, calibration and division of the atmosphere that became possible with the spatial expansion of SCCB devices belies a scientific epistemology of space. This spatial epistemology, which is characteristic of various forms of governmental practice and geographical enquiry, is based upon a belief in a geographical order of knowledge. This epistemological conviction moves beyond a realisation that knowledge has a geographical form, to suggest that space provides preordained frameworks through which knowledge can be excavated. To this end space becomes an epistemological category: a basis for knowledge production. This geographical vision of knowledge is often related to a positivist commitment to the privileged role of scientific methods in uncovering the spatial patternings of truth (see Dixon & Jones, 1996). The methods of spatial science are, however, synonymous not only with the geographical demarcation of knowledge, but also with the careful study of the connections that bind together a place in order for it to be differentiated from other locations. Increasingly in Britain of the 1930s and 1940s we see accumulated atmospheric knowledge being combined with other forms of commensurate geographical data (including economic activity, housing types and fuel use). It was in this context that the 'causal qualities' attributed to the spatial organisation of the world began to provide a basis for the rationalities of air government (see Huxley, 2007: 194).

The work of the DSIR and SCCB in expanding and intensifying air pollution surveillance throughout Britain not only generated a more complete picture of atmospheric pollution, but also enabled new governmental imaginations of the air to emerge. In essence the activities of the SCCB represent the first large-scale geo-coding of the atmosphere produced in Britain. Pickles (2004: Chapter 1) describes geo-coding as the process whereby things of all kinds (people, resources, animals, air) are organised through a common system of spatial reference. The geo-coding of air pollution essentially involves attributing fragments of the atmosphere to specified geographical locations. This process of geo-coding is important because while it may be impossible to govern the atmosphere as a system in and of itself, it is feasible to govern the atmosphere by governing space. The spatialisation of the atmosphere is an accepted method by which air pollution is understood and governed in the world today. This book asserts, however, that the association between atmosphere, government and space was only able to begin in Britain on a large scale during the 1930s.

But the new capacities for air government initiated by the work of the SCCB came with a cost. As previously discussed, one of the problems associated with spatially extensive systems of environmental monitoring is the pressure it places on networks of technological supply, replication and repair. As Chapter Five revealed, even the relatively small-scale monitoring network associated with the CIAP had problems coping with the difficulties of maintaining a fully operational network of atmospheric surveillance. But now with hundreds, as opposed to tens, of monitoring stations the predicament intensified. Space really worked against the SCCB in two broad ways. First, with more and more sites to provide with standard equipment and analysis devices, the supply chains of governmental technology become longer and more expensive to maintain. As discussed in Chapter Five, the nature of atmospheric monitoring meant that equipment was particularly in need of constant repair and replacement. As deposit gauges, filters and trays had to be left out in open spaces for long periods of time they were subject to a range of threats. While polluted air and rainfall tended to take their own toll on the apparatus, there are numerous accounts of local monitoring stations losing devices to frost damage.[33] In addition to environmental damage records also reveal that many gauges and filters were lost in undergrowth and never found again, or their precise location forgotten when monitoring responsibilities moved from one member of staff to another.[34]

In addition to environmental damage there is also much evidence of human error contributing to loss and damage in the SCCB network. Numerous examples of the role of human error in upsetting the monitoring capacities of the SCCB can be found in letters that were sent to the chief officers of the SCCB pleading for new equipment to be supplied to a locality.[35] In regular correspondence was John Edwards of the London County Council. As a scientist with responsibility for supporting and analysing London's network of monitoring devices, it appears that Mr Edwards was well aware of the problems of maintaining an extended network of standard monitoring equipment. On 2 October 1939 John Edwards describes one of the types of accident that could happen in the field,

> I am very sorry to advise you that while trying to place two of the glass bowls in a cardboard box at Lewington they were broken. I alone am responsible. As I raised one bowl by the neck to place it in the box the rubber tube adhering to the stem slipped in my hand and the bowl crashed down on the bowl below.[36]

The necessary day-to-day maintenance of the SCCB's surveillance network clearly increased the probability that accidents such as this would occur on a regular basis. While taking full responsibility for the simultaneous destruction of two sampling bowls, Mr Edwards shifts the blame for another breakage to a surprising source,

> I have enquired as to the fate of the spare bowl housed in the basement of the Newington office [...] The glass [bowl] had been placed by me in a straw-lined cardboard carton and was stored on a wide shelf in the basement. Members of glass-room staff had occasion to move the cartons in the process of cleaning

up the basement presumably in preparation for an air-raid shelter – and when they approached the carton containing the glass bowl a cat leapt out from the bowl, the carton overbalanced and the bowl crashed to the floor and was broken.[37]

On other occasions there are reports of members of the London County Council driving over pollution monitoring equipment in cars and even leaving devices on the roofs of vehicles as they drove away from monitoring sites.[38] While such accidents (although perhaps not of the feline variety) are to some degree inevitable, they did undermine the ability of the SCCB to provide a continuous record of air pollution. With finances tight, equipment expensive and the distances between the DSIR and monitoring sites growing all the time, significant delays would emerge between the loss or damage of a piece of equipment and its replacement. Numerous traces of these delays can be seen in the records of air pollution during the 1930s and 1940s. Many graphs of air pollution produced by inspectors at the time are interspersed with annotations reading 'samples lost' or 'bottle broken'. While such discontinuities in monitoring records may seem innocuous, when replicated across 500 devices nationwide they had significant implications for the comparative analysis of air pollution levels and events desired by government institutions.

The second set of spatial problems that emerged in relation to the SCCB's expanded geographical network of air monitoring relates to the increasing need to use unprotected public spaces to support air surveillance. As discussed in Chapter Five, the relatively small scale of the CIAP's early air pollution monitoring network meant that atmospheric sampling stations could be located on sites owned and controlled by partner organisations (including local authority buildings, hospitals and universities). But with the intensification of surveillance promoted during the 1930s a new range of sites and locations needed to be procured for monitoring activities. Suddenly roadsides and public parks were being used to site deposit gauges and trays, while walls and pavements were exploited as locations for automatic filters and cylinders. While the use of these new sites greatly enhanced the geographical sensitivity of the SCCB's work, such unprotected sites did subject the monitoring equipment to a series of corrupting influences. In London, for example, records of the Public Control Department show that air monitoring equipment was often vandalised or stolen.[39] In his reflections on the work of the DSIR during the 1930s Brimblecombe notes a particularly unusual form of contaminate that afflicted deposit gauges deployed in the public spaces of Glasgow: namely elevated levels of beer sugars (1987: 151). At other urban sampling sites gauges were also corrupted by traces of urine within the collecting bowls.[40] Despite offering an unplanned public service for late night drinkers, the regular contamination and corruption of sampling equipment was clearly a spatial cost of intensifying air pollution surveillance in Britain.

Air Pollution and the British Military Establishment: Vertical Territorialities and the Second World War

Despite the significant barriers that existed to the spatial extension of atmospheric pollution surveillance outlined in the previous section, nothing threatened the maintenance of the SCCB monitoring network to the same extent as the Second World War. As more and more government time and resources were devoted to the emerging war effort, concern with air pollution and support for the work of the SCCB would naturally decline. Suddenly DSIR resources were being directed towards the development of military technologies, while local authorities were focusing on issues of civil defence and safety. Despite these new pressures, records show that many air pollution devices remained serviced and operational throughout the war.[41] Notwithstanding the continuation of air pollution surveillance in many SCCB areas, the war years did see a significant contraction in both the spatial extent and intensity of atmospheric monitoring activities. Not only did the surveillance of atmospheric pollution decline during the war, but there is also suggestion – apocryphal perhaps – of a complete reversal of government policy on air pollution. Whether heightened levels of air pollution were actively encouraged by governmental officials or not, it is clear that smoke was a useful tool of military defence during the air raids that marked the early years of war in Britain. Despite changing levels of state commitment to air pollution issues, however, it is clear that the Second World War did bring with it a series of new developments within the governmental understandings and comprehension of atmospheric pollution.

The fog of war: relocating atmospheric surveillance

It would be erroneous to assume that it was only with the onset of the Second World War that the British military establishment took a scientific interest in air pollution. During its early years, the CIAP had come under the jurisdiction of the Air Ministry, which also had operational responsibility for military air defence and strategy. Furthermore, in the 1920s the DSIR had enlisted the support of J.D. Fry to support the work of its newly formed Atmospheric Pollution Research Committee and Standing Conference.[42] J.D. Fry was superintendent of the Physics Department at Porton, which served the Chemical Warfare Research Department, and had specialist knowledge of the science of atmospheric contamination.[43]

In relation to these institutional contexts, there had been many opportunities for interconnections to be forged between air pollution study and military research. During the Second World War, however, military interests did not only overlap with air pollution study, they also started to shape and transform it.

One of the most obvious ways in which military interests reshaped air pollution study during the war was in relation to its spatiality. In 1939 Bomber Command called for a rapid and continued assessment of fog-related problems and how they were affecting the functioning of aerodromes in Britain.[44] This command would lead to the development of a systematic and continuous regime for the monitoring and assessment of atmospheric conditions in the vicinity of aerodromes. Aerodromes had been sites for meteorological analysis and atmospheric observation long before the onset of the war. What changed from 1939 onwards, however, was the desire to move beyond merely predicting bad weather and fog, to understanding the role of human-induced pollution in contributing to poor visibility at military air bases. In times of extreme military conflict it was not enough merely to cancel flights and sorties because of poor visibility around aerodromes: such a decision could have handed significant strategic advantage to the enemy. Consequently, by calibrating atmospheric observations at aerodromes, with the concurrent surveillance of air pollution in population centres by the DSIR, it was hoped that more could be done to control human-induced smog.

Vertical territoriality and aerological diagrams

The onset of military conflict in Britain not only led to the partial re-spatialisation of air pollution surveillance; it also consolidated a new way of perceiving atmospheric pollution. This re-spatialisation process was in part based on a concern with the impacts of air pollution on strategic military instillations, but it was also based upon a new interest in the nature of atmospheric pollution at different altitudes. The elevated perspectives offered by rising levels of civil and military aviation during the 1930s provided unique insights into the nature, extent and distribution of air pollution. The following excerpt from the *Abingdon Local Weather Phenomena Book* reveals the types of sights that military pilots would often encounter,

> The A.O.C. and S.A.S.O. and Station Commander went to Duxford in the afternoon when the wind was northwesterly 10 m.p.h. Just west of Royston they ran into a lane of smoke and industrial pollution from the Black Country which extended beyond Duxford. They could smell the smoke which was more than 2000 feet thick and visibility in it was reduced to about 2000 yards.

Wing Commander Cumming ran into the same stream of pollution in the morning. He stated that it was about 25 to 30 miles wide and 4000 feet thick. At 400 feet it was beginning to thin out considerably.[45]

While the use of weather balloons and other meteorological equipment had enabled the assessment of the nature of the atmosphere at different altitudes for some time, it is clear that air travel facilitated a new scoptic regime for air pollution monitoring. Air travel meant that observers could now pass in, through and over palls of air pollution. The ability to look down on large swathes of industrial air pollution meant that pilots could assess the extent of air pollution in ways that would have been simply unimaginable to the nineteenth-century smoke inspector. In this particular account, for example, it is clear that air pollution from the Black Country conurbation was polluting up to 30 miles of surrounding land. While the ground-based human and technological monitoring of atmospheric pollution had been able to give accurate measures of the relative severity of pollution events, it was hard to provide reliable accounts of the spatial extent of pollution clouds in this way. Perhaps more important than assessing the spatial extent of pollution, the nature of air travel also enabled pilots to pass through smog and assess its vertical properties. In this record from the *Abingdon Weather Phenomena Book*, Wing Commander Cumming was able to determine the thickness of the pollution cloud enveloping the Black Country. Air travel essentially made it easier to accurately assess the vertical extent of air pollution. While concern with the thickness of pollution had previously been of little interest to urban reformers and air pollution scientists, during the 1930s and 1940s the novel vertical perspective on atmospheric pollution would have a profound affect on the governmental assessment and measurement of pollution in the future.

With the onset of war the study of the vertical extent and distribution of air pollution moved from being a question of scientific intrigue to being of high strategic military importance. It was in this strategic context that the British Air Force used its significant resources and technologies in order to support a new era of high altitude pollution study. At the centre of this new regime of study was a desire to understand the relative distribution of air pollution at different altitudes, the primary factors that caused such distributions, and what could be done to mitigate the threats of atmospheric pollution to air travel. In part this work built upon previous studies carried out in the USA during the 1920s.[46] These studies had sought to determine the relationship between dust concentration in the atmosphere, altitude and diurnal/seasonal variations. British military concern with the distribution of dust particles in the atmosphere related not only to the direct effects of pollution on air transport, but also to the potential role of pollution particles in the formation of moisture and clouds at strategic altitudes.

To this end, studies of the atmospheric concentration of dust particles in the atmosphere were combined with a series of upper air descent tests which were used to analyse the relationship between height, temperature and wind speed.[47]

While the study of the vertical atmospheric distribution of temperature, wind speed and dust conducted by the British military did not constitute a systematic study of air pollution at different altitudes, the knowledge that was gleaned from these studies would provide a new analytical framework within which to interpret the production and distribution of pollution. Through the production of new aerological diagrams of the atmosphere it became possible to understand the conditions under which certain forms of pollution would be produced at certain heights and the likelihood of pollution events spreading over larger spatial areas. Drawing on long-established connections with the Meteorological Office (through the Air Ministry), the Royal Air Force was able to use research into the science of cloud formation to provide predictions of the likely impact of air pollution on the production of wider smogs at certain altitudes and in certain places.

Although the analysis of air pollution conducted by the British military establishment was not concerned with the same types of governmental questions that had informed the emergence of air pollution science in Britain, this chapter claims that military research had a profound impact on the governmental and scientific understanding of air pollution in the decades that followed the war. First, and perhaps foremost, the ability to study the verticality of air pollution meant that a new, three-dimensional understanding of the nature of air pollution would take hold. In essence this shift in perspective on air pollution reflects a move from a flat governmental ontology (represented in the numerous maps and spatial statistics gathered by the CIAP, ACAP and SCCB), to a more vertically nuanced basis for calculating pollution.[48] Second, it is also clear that the elevated perspective on atmospheric pollution facilitated by air travel generated an increasingly acute sense of the spatial extent of air pollution. As the following section outlines, a sense of both the vertical and horizontal nature of air pollution would critically shape governmental rationality towards atmospheric monitoring in following decades.

Space, Volume and the Calculation of the British Atmosphere

At the beginning of this chapter we discussed emerging work on the links between geography, governmentality and space. This body of work has begun to reveal the implication of spatial rationality and sciences of space within different governmental projects.[49] While we have already discussed how a belief in the 'causal qualities of space'[50] informed the early work of the ACAP, it is clear that the 1950s and 1960s saw a new rationality of space starting to shape British air pollution government. Drawing on geometric visions of space, these new forms of air government sought to use various forms of spatial grid to govern the atmosphere. This novel rationality of space – which was based upon the construction of standard three-dimensional measures of the air and atmospheric regions – served both as an ordering device for the production of air knowledge, but also as a paradigm against which to test governmental policy towards the atmosphere.[51]

Perhaps the clearest example of this type of relationship between government and space can be found in Matthew Hannah's work on governmentality, territory and the census in nineteenth-century America (Hannah, 2000). Hannah reflects upon the role of abstract spatial creations, such as the grid, in the geometric framing and subsequent production of spatial knowledge sets. According to Hannah, 'It is impossible to undertake an accurate census unless there is some geographical framework on which to define and precisely locate enumeration districts which exhaust the territory (without overlap) (ibid.: 118).' In addition to preventing the wasteful and confusing overlap of knowledge gathering, abstract spatial registers provide an arbitrary, but nonetheless vital, fixed frame of reference within which to record highly mobile things such as populations, water resource and atmospheres. Despite the highly artificial nature of geometric spatial registers, Hannah also observes that, 'reference grids afford epistemological control of a territory not merely in the abstract, symbolic sense of enabling government to "survey the realm," but also in the sense of facilitating the physical access by the agents of governmental knowledge' (ibid.: 120). In other words, not only can imposed spatial frameworks help to generate useful data sets for governmental perusal, but they also offer invaluable guides to the work of scientists responsible for knowledge production on the ground, as they try to effectively locate their work.

Rethinking atmospheric government and the London fog disaster

While the aerological diagrams of wartime Britain provided the theoretical foundations for a new geo-coding of the British atmosphere, the impetus to implement this new measurement regime came in December 1952. In the immediate post-war period the air pollution monitoring network of Britain was gradually re-established to a level similar to that of its pre-war status. The DSIR's monitoring network was augmented in 1951 when the British Standards Institute issued British Standard 1747.[52] This Standard established national criteria for the form and use of deposit gauges. In December 1952 these nationally sanctioned deposit gauges recorded one of the worst pollution events to ever hit Britain. The 1952 London fog disaster, as it is now commonly called, was so severe that many of the city's air pollution monitoring devices clogged up and became filled to capacity. So much has been written about the London fog disaster that the shocking nature of the event has been lost within a sea of over-determination.[53] However, perhaps more than any other environmental event in British history the disaster revealed the potential impact of air pollution on the functioning of Britain's social economy.

In December 1952 climatic conditions conspired with London's anthropogenically produced air pollution to generate a dense fog that covered more than 1000 square miles (Thorsheim, 2006: 151). The fog lasted for five days leaving large areas of the English southeast with zero visibility and freezing temperatures. The effects of the fog on London's transport systems were widespread and highly debilitating. All air traffic from the capital was diverted to fog-clear airports in other parts of the country (*The Times*, 1952a). Shipping along the Thames was brought to a complete halt; all except three of London's public bus and trolley services were cancelled; and the majority of local and long-distance train services operating into and out of the capital were stopped (ibid.). The situation for private road transport was no better. According to *The Times*, the London Ambulance Service received 334 emergency calls on the evening of Saturday 6 December – 100 times greater than was normal for a Saturday night (ibid.). A large number of these calls were made in relation to the numerous traffic accidents that the fog contributed to. In one incident in Kent 14 vehicles were involved in a huge pile up (*The Times*, 1952b) (Figure 6.2). The radio-controlled patrol units of the Automobile Association claimed that they were having great difficulty finding members who had phoned for help, while AA drivers claimed that there was not a half-mile section of road in the capital where it was possible to see for further than five yards (*The Times*, 1952a). With overland transport of all kinds severely disrupted, most Londoners turned to the underground train network as an alternative. This brought its own set of problems. It was reported that at one Central Line Station at Stratford (east London) 3000 people were queuing to purchase tickets after work: it was claimed that the queue stretched for hundreds of yards (*The Times*, 1952b). The disruption in transport had significant impacts for the social economy of London. With many people not being able to get to their places of work key services and industrial sectors had their activities scaled down. It was also difficult for London to effectively receive and distribute basic resources like food, milk and fuel.

Beyond the immediate economic implications, it was the social costs of the event that perhaps had the most enduring impact on people at the time. Scotland Yard reported that under the London fog various forms of crime, and in particular burglaries, rose sharply (*The Times*, 1952a). In one spate of

Figure 6.2 A London police officer directing traffic during the smog of December 1952
Source: reproduced with permission from Hulton-Deutsch Collection/CORBIS. Corbis image code: HU031865

burglaries it was reported that the same thief entered three flats in Princes Gate via a drainpipe, while others used a ladder to enter a house in Chelsea (ibid.). On Saturday 6 December there was also an unusual amount of street attacks recorded throughout London, with the assailants utilising the fog to secretly approach their victims (ibid.). While the breakdown of law and order in the metropolis was a source of major anxiety, it was the health effects of the fog that would become the most keenly debated and concerning aspect of the entire disaster. As with many large-scale pollution events it is difficult to ascertain precisely the number of health problems and deaths that were a direct consequence of the fog disaster. Dr G.F. Abercrombie, Chair of the Emergency Bed Service Committee of King Edward's Hospital Fund for London, estimated that during the fog cases of respiratory disease increased four-fold, while cases of heart disease were running at three times the normal rate (*The Times*, 1953a). Initial estimates of the deaths resulting from the fog episode ranged from 2,851 up to 6,000 people (Thorsheim, 2006: 162). A 1953 report by Dr J.A. Scott, the County Medical Officer of Health and School Medical Officer of London County Council, claimed that the fog disaster had added an extra 0.5 deaths per thousand in one week for London, raising the death rate to a level similar to that experienced in the influenza epidemic that hit the city in 1918 (*The Times*, 1953b).

Contemporary estimates place the number of deaths caused by the disaster at approximately 4000.[54] The varied impacts of the London fog disaster had a profound effect on government thinking concerning air pollution and its abatement.[55]

Returning to Foucault, if government is all about ensuring the proper disposition of things within tolerable bandwidths of existence (see Chapter 2), it is clear that the events of December 1952 revealed just how serious a polluted atmosphere could be to Britain's socioeconomic order. It was in this context that the public effects of air pollution that were felt on the streets of London in 1952 soon started to reverberate within the halls of Westminster. By mid-December the London fog disaster had become a topic for heated debate within the House of Commons. Initial political discussion about the London fog focused upon the actual number of deaths caused by the disaster. With the Ministry for Health unable to provide reliable estimates of fog fatalities, concerns were raised that the event went beyond existing governmental capabilities of and for air government (Thorsheim, 2006: 159). It was in this context that many Members of Parliament started to call for the formation of a special committee of government to explore in a systematic way the causes and effects of the fog disaster, and to recommend the best course for future governmental action (ibid.).[56] It is interesting to note that calls for a special governmental committee were initially resisted on the grounds that the existing Atmospheric Pollution Research Committee (of the Fuel Research Board) already served that function. As discussed previously, the Atmospheric Pollution Research Committee had been established at the same time as the SCCB (at the moment when the ACAP had become a part of the DSIR). Under extreme Parliamentary scrutiny it became clear that the Atmospheric Pollution Research Committee was not fit for purpose. Quite apart from the fact that the ministerial responses to questions in Parliament clearly revealed that many leading politicians did not actually understand the role of the Atmospheric Pollution Research Committee, many argued that it did not contain the necessary cross-section of expertise to deal with the complexities of air pollution study.[57] Towards the end of January 1953, Dr E.T. Wilkins (the officer then in charge of air pollution research at the DSIR) was able to confirm that the unusually high number of fog-related deaths in the previous months were not the product of any *abnormal poison* in the atmosphere (*The Times*, 1953c) (see Figure 6.3).[58] However, when it was discovered that the Atmospheric Pollution Research Committee had only met twice during the whole of 1952 the government was forced to declare its intention to establish a special committee to investigate air pollution.

In July 1953 the British government announced that Sir Hugh Beaver would head a committee to study the causes of air pollution and strategies for its effective mitigation. Sir Hugh Beaver was a prominent industrialist and

Figure 6.3   An un-named analyst employed by the government's Atmospheric Pollution Research Committee reveals the levels of pollution that were afflicting London in the 1950s. Here the assistant is revealing the air pollution deposits left on the filter of an air-conditioning plant in 1954. The London fog disaster brought the work of the Atmospheric Pollution Research Committee under critical scrutiny
Source: reproduced with permission from the Hulton-Deutsch Collection/CORBIS. Corbis image code: HU020344

engineer and his committee was made up of chemists, medical experts, industrialists and representatives from the Alkali Inspectorate.[59] In this context, as with many of the special committees and commissions that have be instigated in Britain to address the problems of air pollution, the Beaver Committee embodied an coming together of mid-twentieth-century air science and government. Interestingly there was one member of the Beaver Committee (officially entitled the *Committee on Air Pollution*) who was an original member of the 1912 Committee for the Investigation of Atmospheric Pollution: the chemist Dr R. Lessing. The Committee was established under the Ministry of Housing and Local Government and given the following terms of reference,

> To examine the nature, causes and effects of air pollution and the efficacy of present preventative measures; to consider what further preventative measure are practicable; and to make recommendations (Ashby & Anderson, 1981: 106).

Working within this set of parameters, the Beaver Committee produced its final report and associated recommendations in November 1954. Perhaps unsurprisingly, given its strong scientific representation, through its diverse recommendations the Beaver Committee sought to bring a degree of scientific precision and certainty to the government of air pollution. At one level, this scientific ethos can be discerned through the way in which the Beaver Committee understood the nature of the air pollution problem. According to Thorsheim (2006), the Beaver Committee utilised its high public profile to finally dispel many of the myths that continued to inform public understandings of the nature of air pollution problems. Just as atmospheric reformers in the nineteenth century had to battle against popular beliefs in miasma, and the idea that air pollution actually found its origins in poorly drained rural spaces, the Beaver Committee sought to undermine the idea that it was only the combination of anthropogenic air pollution and poor climatic conditions that threatened human health (ibid.: 173–6). Consequently, while it was clear that the London fog disaster was an imbroglio of culture and nature, air pollution and meteorology, economic process and temperature inversion, the elevated death toll would not have existed without the production of harmful atmospheric pollution. Sir Hugh Beaver consequently stated,

> We expressly avoided basing our arguments on the danger to heath of particular incidents, such as the London smog of 1952. Not that we minimised the catastrophe in anyway, but we felt that undue emphasis on it would distract attention from the fact that damage to heath and danger to life were going on all over the country, all the time, year in and year out.[60]

It was in this way that the Beaver Committee was able to utilise the London fog disaster as the basis for asserting what air pollution scientists had known for some time: that ambient air pollution of all kinds represented an ongoing, if often invisible, threat to human health and well-being (ibid.: 166). It consequently recommended that action be taken to mitigate all forms of air pollution and not merely extreme events. The adoption of this governmental position on air pollution also removed the need to prove that smoke pollution was an actual nuisance before a prosecution could be pursued (see Ashby & Anderson, 1981: 107).

The second context within which the Beaver Committee sought to impart atmospheric government in Britain with a degree of scientific certitude came in the way it presented its solutions to the air pollution crisis. According to Ashby and Anderson, the Beaver Committee used its power and influence to assert that the clean air technologies and fuels necessary for strong anti-air pollution legislation already existed in Britain (ibid.: 106–11). It was on the basis of faith in clean air technologies that the Beaver Committee recommended much more powerful forms of atmospheric legislation than politicians had been previously willing to entertain. The key recommendations of the Beaver Committee were that all forms of dark smoke production should be outlawed, and that for the first time this form of air pollution legislation should be applied to domestic as well as industrial premises.[61] An additional, but nonetheless significant, recommendation of the Committee was that local authorities should have the ability to designate spatial areas as *smokeless zones* and *smoke control areas*.

The various recommendations of the Beaver Committee provided the foundations for the now famous 1956 Clean Air Act.[62] According to many environmental historians, Britain's Clean Air Act was the first piece of explicitly environmental legislation passed by any State in the world. While such a claim appears to require a degree of semantic contortion, there is no doubt that the Clean Air Act ushered in a more confident era of atmospheric government in Britain (McNeill, 2000: 64–71; Sheail, 2002). An important component of the Clean Air Act was that it was specifically designed to replace the cumulative, but ultimately confusing, forms of atmospheric legislation contained in previous public health acts. The Clean Air Act also proffered a more definitive definition of the forms of air pollution that it was possible to legally prosecute,[63] while harmonising the regulatory responsibility for enforcing air pollution legislation between central and local government. The overt governmental confidence of the Act was, however, expressed most clearly in the enforcement of smokeless zones and smoke control areas. In many ways these represented crucial shifts in the tactics of atmospheric governmentality.[64] Suddenly the governance of air pollution was not only about the disciplinary surveillance and assessment of whether pollution crossed key thresholds of social tolerability, but instead involved the sovereign expulsion of polluting activities from certain spaces within a city.

Re-measuring British air pollution: geo-coding the air and the National Air Pollution Survey

By the late 1950s there existed both the scientific imagination and the political will to begin to conceive and measure the atmosphere, and associated forms of pollution, in new ways. As argued previously in this chapter, this new regime of air measurement involved an attempt to geo-code the British atmosphere. This geo-coding, which began in the late 1950s, operated on two distinct scales. At the first level, the 1950s witnessed the instigation of the first nationally designated measure for air pollution. The *British standard measure of air pollution* established for the first time both the method and units for measuring pollution in the British atmosphere.[65] According to the standard, air pollution was to be measured on the following basis,

A sample of air drawn for a 24-hour period through a filter paper and the staining measured by a reflectometer. A calibration curve is then used to give the smoke concentration in terms of microgrammes of equivalent standard smoke per cubic metre of air [...] after passing the smoke filter, the sample of air is bubbled through dilute hydrogen peroxide which oxidises any sulphur dioxide to sulphuric acid which is determined by titration to pH 4.5 (Warren Spring Laboratory, 1972a: 4).

While this standard method clearly draws inspiration from the technological advances pioneered by the CIAP (see Chapter Five), what is most significant about this measure is the way in which it sought to frame the calculation of air pollution within a volumetric frame of reference. The principle of measuring the atmosphere in cubic metres indicates a new rationality within the calculation of pollution. Suddenly air pollution was not simply measured in terms of quantity, but in relation to a spatial frame of calculative assessment. While the atmosphere is not, of course, composed of discrete cubic blocks of air, the British standard measure of air pollution reflected a desire to regiment air measurement while acknowledging the volumetric ontology of atmospheric existence. It was not, however, the fact that the British atmosphere was being measured volumetrically that is most significant here, but that an arbitrary volumetric grid was starting to be imagined by State authorities in and through which the atmosphere could be brought under some degree of governmental order.

The second key strategy in the geo-coding of the British atmosphere that began to emerge in the 1950s was a move towards the first national survey of air pollution. Following the London fog disaster and the passing of the Clean Air Act, it was felt by many key figures within government that a national survey of air pollution was required both to assess the efficacy of the Act and to provide a basis for future government policy development. By 1957 Britain had one of the largest national systems of air pollution monitoring in the world with 333 sites dedicated to the daily monitoring of smoke and sulphur dioxide, 993 locations observing monthly levels of dust and grit fall, and 1115 devices monitoring the rate of reaction of atmospheric sulphur dioxide and lead dioxides (Warren Spring Laboratory, 1972a). Significantly, in relation to this chapter, the move towards a national atmospheric census was not based so much on extending atmospheric monitoring capacity, but on the spatial calibration of a national atmosphere.

With a feeling that the existing atmospheric monitoring network was 'rough and ready' and too 'blunt a tool' to deal with new governmental demands for air knowledge, a special Working Group of the SCCB was established in 1959 to explore the best way to conduct a national census of the atmosphere. The Working Group was composed of representatives from a range of relevant government departments with different atmospheric

Table 6.1　Members of the SCCB's Working Group of the National Air Pollution Survey

Chairman:
S.H. Clarke (*Director of the Warren Spring Laboratory*)
Secretary:
Mrs M.-L.P.M. Weatherley (*Warren Spring Laboratory*)
Working Group members:
N. Bastable (*Chair of SCCB and Chief Public Health Inspector, Barking*)
S.G. Burgess (*Chief Scientific Advisory, London County Council*)
S.R. Craxford (*Warren Spring Laboratory*)
L.E. Hockin (*Alkali Inspectorate, Ministry of Housing and Local Government*)
A.R. Atherton (replaced by G. Hopkinson) (*DSIR*)
P.J. Lawther (*Director of Air Pollution Research Unit of the Medical Research Council at St Bartholomew's Hospital*)
P.J. Meade (*Meteorological Office*)
P.J. Harrop (replaced by K.F. Munn) (*Ministry of Housing and Local Government*)
D.W. Slimming (*Warren Spring Laboratory*)

Source: Warren Spring Laboratory (1972a: 3)

remits (see Table 6.1). The Working Group made its report in 1960 and recommended a national survey of smoke and sulphur dioxide that would be conducted by augmenting the existing SCCB monitoring network while deploying the British standard measure of air pollution. The objectives of the survey were described later in the following terms,

1. To provide guidance to central and local government in the application of existing clean air legislation.
2. To assess improvements that are occurring as a result of such legislation or other causes.
3. To provide a technical basis for further legislation if such legislation should be necessary.
4. To provide a systematic body of data for use, by the medical authorities, for epidemiological studies of the effects of air pollution on health, and by universities and government laboratories, and any others who may be interested, for investigations of the effects of weather, urban structure, topography, etc., on the distribution of pollution within towns and of drift from them (ibid.: 4).

While the governmental utility of the National Air Pollution Survey was not significantly different from the motivations behind previous atmospheric monitoring regimes, its geo-coding of the atmosphere was.

Table 6.2 Number of towns sampled of designated size within the National Air Pollution Survey up to end of March 1966

Population range (thousands)	Total number of towns in the UK	Number of towns in the survey
Above 100	62	59
50–100	102	77
20–50	252	122
5–20	399	94

Source: Warren Spring Laboratory (1972a)

The National Air Pollution Survey had two key implications for the geographies of air pollution monitoring in Britain. First, the survey involved an inevitable spatial expansion in the use of monitoring equipment sanctioned by the SCCB. This expansion process provided the SCCB with the opportunity to think strategically about where to locate its new monitoring devices. There was much discussion within the Working Group over the best spatial formula to use for location-based decisions and which strategies would best support a representative atmospheric sample. Ultimately, the Secretary of the Working Group, a Mrs M.-L.P.M Weatherley, developed an effective formula for determining the best location of new monitoring stations. Weatherley's spatial system was based upon a sampling scheme devised by Professor W.B. Fisher of Durham University and sought to ensure,

> A fair distribution of towns [...] taking into account their population, population density, domestic heating habits, industrial and other activities and also their ventilation characteristics, since a given town in an enclosed valley will present very different air pollution problems from another, otherwise identical, town situated on a windy upland (ibid.: 5).

Such sampling criteria led to 353 towns being incorporated into the National Air Pollution Survey with an additional 200 rural locations also being utilised (see Table 6.2).[66] The second change in the geo-coding of the atmosphere evident was the spatial presentation of the data. The National Survey ran between 1961 and 1971. The government's Warren Spring Laboratory was responsible for collecting and collating the data that was collected through the sample sites. The Warren Spring Laboratory produced a series of regular atmospheric pollution bulletins containing the raw data being gathered through the survey, however, the main results were published in a five-volume report in 1972. While the bulletins presented air pollution data on a station-by-station basis, the final report organised results on the basis of regional grids.

To this end, the nationalisation of air pollution monitoring in Britain was directed more towards the regional spatialisation of the atmosphere than to the development of a national picture of air quality. These regional atmospheric registers are significant because they embody the first attempt to represent the British atmosphere in explicitly spatial terms. The British atmosphere was divided into 12 air regions (including the Greater London Council Area) (see Figure 6.4). The spatial parameters of the regions that were used by the Warren Spring Laboratory were chosen to deliberately mimic those deployed by the Registrar General. The use of these statistical regions was significant in terms of atmospheric government because data sets on fuel consumption, population size, housing types and industrial infrastructure already existed for these discrete spaces.

In the final five-volume report produced by the Warren Spring Laboratory key national trends in air pollution were noted (ibid.: 11–14). Overall a significant steady decline was observed in the emissions and concentration of smoke and sulphur dioxide across the country as a whole. The remainder of the final report was, however, dedicated to presenting pollution statistics on the basis of regional spaces. While reports on individual regions were often presented on a town-by-town basis, aggregate regional figures provided a picture of British atmospheric pollution that was highly legible to central government departments (see Table 6.3).[67] The trend diagrams of regional atmospheric pollution enabled central government to compare the relative rate of success of regions in reducing different forms of air pollution. One of the most striking spatial comparisons facilitated by the National Survey was the relative levels of smoke emissions in the northern and southern regions of Britain. It was estimated that smoke production in northern regions was approximately three times higher than that of their southern counterparts (this pattern was largely attributed to much higher rates of domestic coal burning in northern regions) (ibid.: 15). Whether such spatial differences were a product of climatic conditions in the north, or the more rapid modernisation of heating systems in the south remained unclear (ibid.), but it is not difficult to see how such spatial data could be deployed to assist in the acts of atmospheric government.

Above all else, the National Air Pollution Survey marked a shift in the spatial matrix of atmospheric knowledge production in Britain. Suddenly, atmospheric knowledge was not only conceived of on the basis of its spatial location, but as part of a broader system of territorial governance (including climates, mineral resources/coal type, housing stocks, population densities and wealth). This transformation also reveals the subtle ebb and flow of the governmentalities of security and discipline within British atmospheric policy. While the activities of the SCCB during the 1950s and 1960s had essentially been dedicated to monitoring the bandwidths of acceptable pollution established by the Clean Air Act, by using political space as a way of

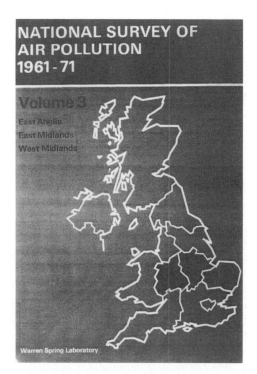

Figure 6.4   Final report of the National Air Pollution Survey, Volume 3 (Warren Spring Laboratory)

Table 6.3   Domestic coal consumption and average smoke concentrations in the atmosphere – National Air Pollution Survey

Region	Domestic coal consumption per head, 1970 (tonnes)	Average smoke concentration (μgm^3)	
		1969–70	*1970–1*
North	0.56	95	88
North West	0.52	90	81
Yorkshire and Humberside	0.50	83	80
Northern Ireland	0.54	80	72
Scotland	0.38	79	69
East Midlands	0.44	64	60
West Midlands	0.34	54	50
East Anglia	0.29	46	46
London	0.04	46	42
South East (excl. London)	0.14	34	31
Wales	0.54	32	33
South West	0.18	31	29

Source: Warren Spring Laboratory (1972a: 14)

aggregating atmospheric data, the National Survey made it possible to think of new forms of governmental intervention (through regional development policies, local authority regulations and educational initiatives) within socio-atmospheric relations. To this end, it is not so much that spatial formulae for knowledge production reflect existing State rationalities, but that calculative spatial projects and rationalities can in themselves inspire changes in the techniques of government.

Conclusion: Governmental Rationality and Spatial Epistemology

Building on the insights of Chapter Five, this chapter has asserted that the command of space is *sine qua non* to the establishment of effective forms of atmospheric knowledge production and government. Following the formation of the Standing Conference of Cooperating Bodies in the 1920s, Britain witnessed the gradual spatial expansion of atmospheric monitoring, as more and more towns initiated air surveillance schemes within their jurisdiction. With the onset of the Second World War the spatial extent of air monitoring diminished, but a new sense of the vertical territoriality of the atmosphere emerged. In the context of military conflict we saw how air pollution took on a new strategic level of significance and how the Royal Air Force supported research into the formation and behaviour of air pollution at different altitudes. Crucially, this chapter asserts that the war years were vital in the movement away from a flat ontology of atmospheric pollution to an increasingly volumetric governmental approach to air studies.

If the 1930s and 1940s were crucial to the spatial expansion of atmospheric monitoring – in both horizontal and vertical terms – this chapter has revealed that that 1950s and 1960s saw the rise of a new system for geocoding the atmosphere. Following the London fog disaster of 1952 and the Clean Air Act of 1956, it became apparent that a new system of atmospheric government and knowledge production was required. In this context the instigation of the British standard measure for air pollution and the National Air Pollution Survey sought to use geometric and territorial spatial frameworks as lenses through which to view, study and govern air pollution. What both the standard measure and National Survey (*inter alia*) ensured was that the atmosphere was not only monitored in space, but also measured by it. To this end, the regional atmospheric registers produced by the National Air Pollution Survey embody the subjugation of the air to space and, by definition, to the potential for governmental discipline. The changing modes of governmental understandings of air pollution that were made possible by these regional registers and attendant spatial rationalities reveal that space is not only a necessary accomplice of governmental strategy implementation, but can also be a powerful calculative paradigm in the shaping of government policy. Yet the question remains as to the long-term effects of such governmental projects on other ways of knowing and interacting with the atmosphere. As the following chapters will illustrate, the spatialisation of the air has remained a popular way of governing the atmosphere in Britain, but one that continues to exercise epistemological restrictions on other kinds of air knowledge.

Chapter Seven

Automating the Air: Atmospheric Simulations and Digital Beings

The National Air Pollution Survey represented the culmination of an almost 100-year struggle to ensure a territorially comprehensive and scientifically rigorous survey of British air pollution. Never before had so many people, over such a wide area, been involved in the simultaneous measurement of a national atmosphere. With hundreds of devices operating in 353 towns and 200 country locations, the National Air Pollution Survey was the crowning achievement of the British governmental science of air pollution monitoring. It is perhaps then with no small amount of irony, that even before the Warren Spring Laboratory had published the final regional reports of the National Air Pollution Survey, the British government was already starting to question the value of such a time-consuming atmospheric endeavour.

There were two broad reasons why the governmental role of the National Air Pollution Survey came under scrutiny in the early 1970s. First, the manual nature of extensive data collection and analysis associated with the National Survey looked clumsy and burdensome in comparison to increasingly sophisticated systems of computer modelling that were becoming available. It was, for example, in 1970 that *The Club of Rome* sponsored the *Project on the Predicament of Mankind* commenced at the Massachusetts Institute of Technology (MIT). Under the directorship of Dennis Meadows the *Project on the Predicament of Mankind* sought to extend the dynamic systems modelling developed at MIT to a global level in order to understand the future implications of population change, resource use and pollution.[1] At a British government meeting to discuss aspects of environmental pollution on 13 July 1971 the Secretary of the State for the Environment consequently reflected,

> The Department of the Environment had an interest and should involve itself in the mathematical modelling approach to population growth, resource

utilisation and pollution [...] It was not enough simply to keep in touch with the Americans – the government would need to put some effort in, if only to win a measure of leverage, and to cover British conditions.[2]

It appears that compared to what was going on at MIT the National Air Pollution Survey already represented a relic of governmental science.

The second reason why doubt started to be cast over the utility of rigorous atmospheric monitoring was geopolitical. In the wake of growing social consciousness of the global threats posed by various forms of pollution, the late 1960s and early 1970s were marked by the first attempts to develop a set of United Nations sponsored agreements on global environmental policy. In the preparatory meetings immediately preceding the United Nations Conference on the Human Environment (convened in Stockholm in 1972), for example, it was noted that the ability to collect and analyse environmental data effectively in the computer age would critically affect Britain's ability to shape international environmental policy.

This chapter begins to chart the role of the twin processes of technological development and geopolitical change in instigating an *environmental revolution* in British government (this process is continued in Chapter Eight) (see Owens & Raynor, 1999). This environmental revolution involved both a change in the technologies that were deployed to monitor environmental systems, and the ways in which the government conceived of pollution threats. While Chapter Eight considers the impact of this revolution on the rise of new forms of ecological governmentalities within British atmospheric government, this chapter focuses on the technological aspects of this upheaval. To these ends, this chapter is distinct from the discussions of technological development and instrumentation conveyed in Chapter Four because it focuses explicitly on the governmental impacts of automated and digital configurations of atmospheric knowledge production. At the centre of this chapter is a desire to illustrate the impacts that automation and digitisation have had on both the science and government of British air pollution. While a significant portion of this project is based upon the suggestion that both automation and digitisation have created unique opportunities for the cyberspatial simulation of atmospheric processes, analysis does not wish to argue that air science and government have somehow been displaced into the artificial worlds of virtual reality. Rather this chapter charts how cyberspatial techniques of scientific knowledge collection, storage and analysis have affected the material practices in and through which States and individuals govern their atmospheric relations.

This chapter begins by reflecting on the critical reviews that were conducted of the National Air Pollution Survey 1961–71, and how these assessments laid the grounds for the gradual formation of supposedly more streamlined and cost-effective systems of 'smart' air monitoring. Analysis then moves on to explore the historical emergence of automated air pollution monitoring and how automated techniques eventually combined with digital technologies to facilitate the growth of atmospheric simulation and real-time governmentalities. The final section considers the impacts of the cyberspatialisation of air pollution knowledge on personal forms of air conduct and being.

Reviewing the National Air Pollution Survey and the 'Environmental Revolution' in British Government

Rethinking atmospheric pollution and the environmental revolution in British government

It is important to position the review of air pollution monitoring conducted in Britain during the late 1960s and early 1970s in relation to broader ideological and institutional changes in pollution policy that occurred at around the same time. The broad significance of this period for State and environment relations in Britain is perhaps conveyed most effectively by Owens and Raynor who have likened this time to an 'environmental revolution in Whitehall' (1999: 7). The changing governmental approach to environmental pollution really began in 1969 when Anthony Crossland established the *Central Unit on Environmental Pollution* (hereafter CUEP). Originally formed as a Cabinet Committee, the CUEP was incorporated into the newly formed Department of the Environment (hereafter DoE) in 1970, and had responsibility for integrating pollution policy across the fledgling department (Jordan, 2000: 12) (The work and remit of the CUEP will be explored in greater detail in Chapter Eight.) The second key component of the environmental revolution in British government was the establishment of the *Royal Commission on Environmental Pollution* (hereafter RCEP). The RCEP met for the first time in 1970 with Sir Eric Ashby (a Cambridge botanist) acting as Chair (Owens & Raynor, 1999: 8). Constituted as an Independent Standing Body, the RCEP brought together a range of experts from beyond the ranks of government including scientists, business leaders, legal experts and engineers. According to the RCEP's Royal Warrant its role is,

> To advise on matters, both national and international, concerning the pollution of the environment; on the adequacy of research in this field; and the future possibilities of danger to the environment. Within this remit the Commission has freedom to consider and advise on any matter it chooses; the government may also request consideration of particular topics.[3]

The intensive auditing of British pollution monitoring and science conducted by the CUEP and the RCEP during the early 1970s had a profound impact on government intervention in environmental affairs. While the RCEP produced regular reports on different aspects of environmental pollution, the CUEP produced a series of so-called *Pollution Papers* (many of which were in direct response to Royal Commission reports – see Department of the Environment, 1975). It is through these reports and papers that the nature of environmental government in Britain started to be gradually recast. In Chapter Eight we consider the impacts of the RCEP and CUEP on governmental thinking concerning environmental policy. In this chapter, however, I am primarily interested in how the new ethos of governmental review produced by these bodies recast the technological basis of air pollution monitoring in Britain and affected the nature of atmospheric knowledge production and government.

The National Air Pollution Survey in critical review: the costs of disciplinary surveillance

It was within the ethos of the assessments of British environmental government instigated by the CUEP and RCEP that a comprehensive review of air pollution surveillance was conducted in the early 1970s. In 1971 the *Interdepartmental Committee on Air Pollution Research* (hereafter ICAPR)[4] convened a series of meetings in order to critically appraise the future of the National Air Pollution Survey (hereafter NAPS). At an early meeting of the ICAPR, convened on 22 November 1971, Dr Robinson of the Warren Spring Laboratory made the following candid observation, 'At present it [the NAPS] constitutes a major part of the Warren Spring Laboratory's work, but it is not highly regarded in the scientific world and does not provide attractive work for scientists'.[5] Dr Robinson's sentiments in part reflect the great difficult that the Warren Spring Laboratory experienced in trying to attract the best and brightest scientists to become involved in the NAPS. The scientific anxieties surrounding the NAPS predominantly derived, however, from the cumbersome temporal and spatial scale of the survey. Members of the ICAPR felt that although the extensive spatial territory and temporal period covered by the NAPS had produced an unprecedented atmospheric data set, the scale of the survey had undermined the scientific and governmental utility of much of the data. As Scott (1998) reminds us, in order to be governmentally useful, data must make that which is to be governed more legible to the State. As we have seen in previous chapters, however, increases in the scale of scientific and governmental surveys routinely result in increases in errors that militate against the legible simplicity desired by States. The scale of the NAPS meant that many gaps existed in the data set it produced. Such spatial and temporal gaps were caused *inter alia* by the loss of data from monitoring stations and the failure to record data over given periods of time. While such lacunas did not have a significant effect on the overall quantity of data collected by the NAPS, they did have a considerable impact on what could be done with the knowledge collected. With incomplete data sets, the calculation of monthly and yearly averages became difficult, while the ability to construct comparable spatial statistics was severely undermined.[6]

At an exploratory meeting convened by the ICAPR to discuss the NAPS, the voluntary nature of the undertaking was identified as a major barrier to the efficacy of the survey. As one member of the Committee noted,

> You will realise that such information is not always available even under the present system. It is the laggard authorities who are least likely to maintain air pollution gauges, and, as you know, we have no power at present to compel them to do so.[7]

Without the political power to compel local authorities to maintain more rigorous records of air pollution data it appeared that much of the great effort required to continue a national census of air pollution would be wasted. It was in the context of such scientific and governmental uncertainties that in 1972 the ICAPR formed the *Working Party on Air Pollution Monitoring*. Chaired by Graham Fuller, the Working Party provided a comprehensive review of the NAPS, conducted an interdepartmental questionnaire of air pollution monitoring needs, and commissioned research into the potential deployment of new air pollution monitoring technologies (research that was conducted by the Warren Spring Laboratory). The final report of the Working Party – commonly referred to as the *Fuller Report* – calculated that only 65% of the monitoring sites that operated in the NAPS provided data that could indicate valid annual averages for air pollution, while only 50% of sites provided the annual averages in consecutive years required for trend analyses.[8] The *Fuller Report* also uncovered the significant periods of time that scientists working at the Warren Spring Laboratory had to spend simply administering the steady flow of atmospheric data from the NAPS. The report claimed that between 80 and 90% of data forms arriving at the Warren Spring Laboratory required some form of clarification (from finding missing site codes to dealing with illegible entries). The administrative costs and difficulties experienced by the Warren Spring Laboratory are typical of the problems associated with extensive governmental knowledge-gathering regimes. Such difficulties are not discussed directly by Foucault. Even in Scott's work, emphasis is very much placed upon the role of statistics and standard measures in making the complexities of the social and environmental worlds legible to governmental authorities. But what if a surfeit of (often less reliable) statistical data actually obfuscates, rather than clarifies, the objects and objectives of government? The collection of more spatially extensive data carries with it the likelihood of greater statistical error and prohibitive administrative costs. Certainly in the case of the NAPS, more knowledge does not necessarily make for better government.

In the context of the governmental failings of the NAPS, the *Fuller Report* considered the benefits of both spatially reduced and temporally discontinuous atmospheric surveys.[9] It is interesting to note that in discussions of the value of reducing the scale of air pollution monitoring in Britain we see the emergence of a tension between the biopolitical and environmental rationalities of atmospheric government. The environmental revolution in British government had in part been inspired by a growing concern with the nonhuman costs of environmental pollution and the need for a more extensive environmental monitoring system (see Chapter Eight). It was in this context that medical experts most vigorously opposed the scaling-back of air pollution monitoring. While it was possible to imagine a scaled-back atmospheric survey that could be redirected to broader environmental concerns, medical officers argued that many of the benefits of the NAPS to health studies would be lost if a less extensive study was instigated. The Chief Medical Advisor on the ICAPR (Dr Martin) defended the medical value of extensive air pollution monitoring. According to Dr Martin, the maintenance of a spatially extensive and temporally continuous record of air pollution offered the greatest opportunity for health scientists to be able to correlate medical statistics with atmospheric conditions and to analyse the links between air quality and a range of medical conditions. A letter from M.W. Holdgate (a member of the ICAPR) reflecting on a discussion held with Dr Martin helps uncover the debates that surrounded the relationship between the NAPS and medical science,

> Dr Martin said that he was opposed to scaling the survey down but his argument appeared to be that scaling down must inevitably reduce the scientific value of the results obtained besides having a political effect on local authorities. Neither of these assertions appears to me to be valid without a supporting analysis. It could well be that higher quality medical information could be obtained from 200 first-class sampling sites [rather] than from 1200 sites whose distribution and range of operations were circumscribed by the nature of the support available from local authorities.[10]

The debate appeared to be between the governmental desire for a smaller number of more easily controlled (and thus reliable) air monitoring sites, and demands for a spatially extended system of air pollution surveillance that could more effectively meet the purported needs of medical science.

In addition to opposition from an influential medical lobby, the Working Group also faced strong governmental resistance to the formation of a diminished atmospheric survey. With many in government still remembering the scientific failings of the State in the wake of the London fog disaster, R.G. Adams wrote to Fuller to caution against any emasculation of the national air pollution monitoring apparatus,

> [t]he proposal would need rather careful promotion in view of current Parliamentary views on the importance of information about smoke concentrations and in view also of the interest felt in the forecast pattern of sulphur dioxide concentrations during the next few years.[11]

It appears that whatever the administrative shortcomings of the NAPS, and the potential scientific value of a smaller scale monitoring apparatus, there was political value in maintaining a costly, but extensive survey bureaucracy. If nothing else, such a bureaucracy would indicate a degree of political commitment to air pollution policy that had been so lacking in the 1950s.

In the context of competing, and often contradictory, scientific and political interests, the *Fuller Report* offered a compromise position on the future of atmospheric pollution monitoring in Britain. It recommended extending the range of air pollutants monitored by the State (particularly in relation to hydrocarbons) and the formation of new 'super sites' for the integrated monitoring of atmospheric pollutants. It also suggested restructuring the NAPS, but supported the maintenance of a wide-ranging monitoring network throughout the UK.[12] A crucial part of this restructuring process was to involve reconsidering the way in which sites were chosen for air monitoring activities. As the RCEP noted,

> The sites used for the National Survey have not been chosen on any systematic basis. The Survey was originally set up by Medical Officers of Health for their own purposes. Each local authority decided whether, and if so, where, monitoring apparatus should be installed and although the results have been centrally coordinated since 1914 there has never been central direction on the sites selected, although advice has been available. The result is that there are areas in the country which probably should carry out monitoring but do not contribute to the survey (Royal Commission on Environmental Pollution, 1976: 87).

Following the recommendations of the *Fuller Report* (and the RCEP), greater powers were afforded to central government to determine the location of air monitoring equipment and to overcome the tyranny of local vagaries in the placing of air surveillance technologies. The *Fuller Report* served to create a governmental space for nurturing and developing a new breed of experimental monitoring technologies: a series of strategically located and integrated automated samplers.

Automation, Digitisation and the Birth of the Real-Time Atmosphere

Continuous air measurement in Britain and the rise of real-time governmentality

One of the main problems associated with the NAPS was that it was labour intensive and thus highly time consuming. The NAPS required the daily collection of filter papers and the protracted analysis of the stains they exhibited. The daily collection of results from different sample sites required different local authority employees to travel over large parts of the towns and cities within which they worked. It was in the context of such bureaucratic inefficiencies that the potential for labour-saving automation started to be explored within British air pollution science during the early 1970s. Automation appeared to offer the promise of less costly and more effective atmospheric monitoring systems envisaged within the *Fuller Report*. While it is temping to assume that the development of automated atmospheric sampling is a relatively recent event, it is possible to trace the automated measurement of air pollution right back to the work of the Committee for the Investigation of Atmospheric Pollution. In the *Report of Observations in the Year 1917–1918* the CIAP (by then renamed the Advisory Committee on Atmospheric Pollution) provided details of a new automatic filter device that had recently been developed.[13]

The automatic device developed by CIAP scientists was, however, complex and cumbersome, and was consequently only used sparingly. It was only following the extensive review of environmental monitoring systems conducted in the late 1960s and early 1970s that the governmental potential of automated air sampling started to be realised. There were only a very small number of automated monitoring sites in existence during the early 1970s. Although the number of automated sites rose to double figures during the mid-1970s – largely in response to the reviews conducted by the Fuller Committee, RCEP and CUEP – it was not until 1987 that we start to see a steady increase in the capacity for automated air monitoring in Britain. The year 1987 witnessed the establishment of the UK urban monitoring network.[14] This network was formed in response to new European Community directives on air quality standards.[15] While the need to comply with the data monitoring demands and standards of an increasingly supranational political space did inform the spread of continuous automated air sampling in the Britain, the growth of the network was also driven by a series of more local political pressures (see Jordan, 2000).

In his fascinating discussion of the *political chemistry* of air pollution monitoring in Britain, Andrew Barry describes how a confluence of environmental and political forces came together in the late 1980s and early 1990s to support the expansion of continuous and automated air monitoring (Barry, 1998: 5, 2001: 153–74). At one level the rise to political prominence of the British Green Party and the lobbying efforts of Friends of the Earth placed pressure on the government to develop a state-of-the-art air monitoring network in the UK (Barry, 2001: 157). Echoing the debates over the NAPS in the early 1970s, criticism of Britain's atmospheric monitoring network was not purely based on a perceived lack of effort by government. Instead there was a concern that despite spending something in region of £5 million per year on air pollution monitoring, Britain was still lagging behind its European neighbours and North American cities in the quality of atmospheric data it was producing (ibid.: 157). In addition to the geopolitical pressures of technological competition, in December 1991, exactly 39 years after the fog disaster, London suffered another costly smog episode (ibid.). During three relatively calm days in December 1991 levels of pollution became dangerously high (in the case of nitrogen dioxide, its concentrations of 423 parts per billion are the highest on record) (Brown, 1994: 4). With severe lung disease up by 22% and the number of people dying from cardiovascular complications up by 12% during the week of the smog, it was estimated that the pollution event was responsible for an extra 160 deaths in the capital (ibid.). The London smog of 1991 raised two crucial issues for atmospheric government in Britain. First, it was realised that the government levels set to determine dangerous thresholds of air pollution were far too high: the permissible limit set by the Department of Health for atmospheric nitrogen dioxide was 600 parts per billion, and yet levels of 423 parts per billion had caused severe problems in London (ibid.). Second, it was recognised that the automated monitoring network operating in the capital was not sophisticated enough to provide an early warning of dangerous air pollution events.

It was in the context of this confluence of pressures and events that in 1992 the Department of the Environment established the Advanced Urban Network (hereafter AUN) to provide a more extensive and sophisticated armature of atmospheric monitoring in British cities (Air Quality Archive, 2008). This new breed of automated sites combined the very latest in air monitoring technologies with a capacity for digital storage and transfer of air pollution data (see Figure 7.1). Two things distinguish modern automated air monitoring devices and stations from their historical predecessors. First, is their ability to detect ever more subtle and transient changes in the composition of the air. Second, is the deployment of digital technology within the automated sampling devices. Unlike the analogue devices that were developed by the CIAP, the ability to make digital records of air pollution meant that atmospheric data was far more mobile than it had been in the past. It is easy for the historian of government and science to overlook the profound significance of digitisation on the appropriation and use of knowledge. In his discussion of the political economy of cyberspace, Timothy Luke provides crucial insights into the impacts of the digital age on government. According to Luke, there is a critical difference between the mobile *bits* of digital data codes, and the more cumbersome atoms that help to convey traditional mediums of data. Luke observes, 'bits can flow effortlessly from cyberspace to cyberspace in milliseconds amidst a building, over the back fence, across a nation, or around the world' (1995: 16). It is the imperceptible speed, and almost infinite mobility, of digital data codes that provide the key to understanding their impacts on both the science and government of air pollution. While I talk at greater length in the next section about the impacts of digital air monitoring on governmental imaginings of the atmosphere, at this point it is crucial to recognise that digitisation is essentially what enabled automated air sampling to reach its full governmental potential: namely atmospheric government in real time.

The application of digital technologies in Britain meant that not only could air pollution be monitored on a continuous basis, but that the results of atmospheric surveillance could be relayed directly to governmental

Figure 7.1   Automated air quality monitoring station, Centenary Square, Birmingham

officials and health officers. While analogue automation meant that the air could be monitored without the regular, daily collection of samples, it still required the physical collection of filter disks on a routine basis. Even with monthly pollution bulletins, it took the NAPS the best part of 11 years to be able to produce its final results. The new breed of monitoring stations, which started to spread throughout Britain during the 1990s, meant that air pollution could be monitored continuously, analysed automatically and relayed to key officials (and more recently the public) without the physical intervention of a scientist, or maintenance engineer, at the site. In the wake of the London smog of 1991, this new capacity for digital air surveillance changed the temporal scope of atmospheric government. Suddenly government officials could issue an air pollution warning in real time. It is in this context that the 1990s represented a key shift in the reasons for air pollution government. Suddenly, the changing science of atmospheric surveillance meant that air government was not merely about assessing pollution events, and their associated effects, once they had happened. Instead, atmospheric government was about actively intervening in order to try and prevent the worst consequences of air pollution actually coming to pass.

The spatial coverage of digital, early warning air monitoring in the UK increased greatly in 1995 as all compatible national and local authority auto-mated stations were amalgamated into a single system (Air Quality Archive, 2008). In 1998 further integration occurred within the automated air sur-veillance network when the UK's urban and rural automated stations became part of the same standardised system of pollution monitoring (ibid.). By 2005 this new *Automated Urban and Rural Network* (hereafter AURN) com-prised 123 monitoring stations: with 87 in urban locations, 22 in rural areas, and 14 compromising the *London Network* (ibid.). Sixty of the sites operating within the AURN are run by national government departments (the Depart-ment for the Environment, Food and Rural Affairs (Defra) in England, and the devolved authorities in Northern Ireland, Scotland and Wales), while local authorities run a further 63 stations (ibid.). The collection of real-time data from these AURN sites is overseen by the non-governmental organisa-tion *AEA Energy and the Environment*, which is based in Oxford.[16] Real-time atmospheric pollution data is relayed by the AEA to Defra's Air Quality Archive website, which can be accessed by the public.

Simulation, surrogate skies and the changing cartographies of air government

If the early 1970s highlighted the potential value of the automated sampling of the air, it is clear that the early 1990s provided the impetus for the devel-opment of an integrated spatial network of automation in Britain. The growth of the AURN has, however, not only affected the temporalities of atmospheric government, but also the way in which it is now possible for governmental authorities to perceive of the spatial form of air pollution. One of the key consequences of automated air sampling was that suddenly the British government had a far more extensive set of data on air pollution than at any previous point in history. When the regular samples of the AURN's 123 stations were combined with the ongoing readings of the government's analogue and passive sampling network (such as the sulphur dioxide monitoring network) a previously unimaginable level of atmospheric surveillance was facilitated. By 2002 it was estimated that over four million air pollution measurements were made in Britain of a range of different chemical pollutants (Air Quality Archive, 2008). In addition to the increasing number of air monitoring samples being produced, the 1980s and 1990s also saw a significant growth in the coordinated collection and calibration of *surrogate data* that could be used to calculate air pollution estimates (see below). The explosion of atmospheric data production, and the fact that much of this data was now available in digital form, provided a new set of opportunities for envisaging and intervening in atmospheric affairs. At the centre of this new field of opportunities was the potential for enhanced computer-based atmospheric pollution modelling and simulation.[17]

The British government's National Atmospheric Emission Inventory (hereafter NAEI) is at the forefront of this new era of air pollution modelling and simulation. The NAEI is run by AEA Energy and Environment on behalf of Defra and the devolved administrations in Northern Ireland, Scotland and Wales. The role of the NAEI is to provide accurate estimates of air pollution and to construct maps of atmospheric emissions to assist in government policy development and strategy. The NAEI is responsible for producing emission estimates and maps for 25 pollutants on a routine basis (see Table 7.1) (King *et al.*, 2006). The NAEI produces atmospheric emulations (predicting the distribution of air pollution) on the basis of two types of data: *reported emissions* and *estimated emissions* (ibid.: 4). Reported emissions are the physically recorded samples of air quality provided by the government's AURN and passive sampling networks. Estimated emissions are derived from a range of surrogate statistics taken from national surveys of energy consumption, transport levels, and even livestock numbers and fertiliser usage (ibid.: 4).[18] These surrogate statistics are converted into emission estimates on the basis of known levels of pollution produced for the average mile travelled by a car, or the operation of a power plant for one hour, depending on the statistics in question (these figures are known as *activity statistics*) (ibid.: 4). This statistical process was described to me in an interview I conducted with a representative of Defra's statistical division in the following terms,

Table 7.1   Emission types mapped by the NAEI

1,3-butadiene	Nitrous oxide (N_2O)
Benzene	Methane
Carbon monoxide	Arsenic
Carbon dioxide	Cadmium
Particulate matter (PM_{10} and PM_{25})	Chromium
Nitrogen oxides	Copper
Non-methane volatile organic compounds (N-MVOC)	Lead
Sulphur dioxide	Mercury
Ammonia	Nickel
Benzopyrene	Selenium
Dioxins	Vanadium
Hydrogen chloride	Zinc

Source: King *et al.*, 2006: 1

Basically, a lot of it is done in terms of working out an emissions factor and then knowing the amount of a certain activity that has been carried out. So per ton of coal burned by a power station you get so much CO_2, you know how much coal this power station is burning from DTI statistics and you multiply the one by the other and that's the source of a large part of Inventory [...] so you need therefore estimates of road traffic, estimates of power station activity, estimates of various other things [...][19]

While the emissions figures produced from surrogate statistics are only general estimates, when combined (and compared) with recordings of actual levels of pollution, at certain sites, they do provide an indication of actual air pollution levels that would be, for purely practical reasons, impossible to develop from an instrument-based network alone.

While surrogate and actual air pollution figures both support the production of national estimates of atmospheric pollution emissions, it is important to recognise the differing governmental utilities of the actual and surrogate statistics marshalled by the NAEI. This distinction was described in the following terms,

There are basically two primary sources [of atmospheric pollution data], a source for emissions and a source for concentrations in the atmosphere [...] the Inventory is all about the emissions as opposed to the concentrations [...][20]

While the NAEI provides the government with an indication of its overall success in meeting pollution reduction targets, the instrument-based measurement of air pollution concentrations provides an insight into the uneven, local impacts of air contamination.

Many of the surrogate statistics detailed above have spatial reference points, with emissions of activities being recorded on an area basis. It is these area statistics, when cross-referenced with point source AURN readings, which enable the production of emissions maps. Crucially the production of accurate, computer-generated emissions maps would not be possible without the level of data density provided by the combination of automated air sampling and surrogate statistics. What is most interesting about the numerous other emissions maps produced by the NAEI is what they can tell us about the links between atmospheric simulation and air government. Care needs to be taken when referring to maps produced by the NAEI as simulations. This analytical care does not derive from a denial that emissions maps are an atmospheric fabrication, but rather from the realisation that even as an overt fabrication of the air they arguably represent one of the most accurate spatial registers of nitrous oxides in the British atmosphere ever produced. Such a realisation raises important questions about the relationship between atmospheric government and atmospheric truth. As we have seen throughout much of this volume, the scale, complexity and perpetual transience of the atmosphere means that air government can never be based on absolute empirical certainty. Atmospheric government requires simulation: it thrives on emulation. In his classic study of the arts and sciences of simulation Baudrillard asserts that 'To simulate is to feign to have what one doesn't have' (1981 [1994]: 3). The emission maps produced by NAEI conform to Baudrillard's definition of simulation to the extent that they embody confident cartographic representations of something the British government does not possess: namely a complete territorial knowledge of the atmospheric concentration of nitrous oxides at any given moment in time. Yet Baudrillard refines his definition of the practices of simulation thus,

> [...] simulating is not pretending: 'Whoever fakes an illness can simply stay in bed and make everyone believe he is ill. Whoever simulates an illness produces in himself some of the symptoms' (Lettré). Therefore, pretending, or dissimulating, leaves the principle of reality intact: the difference is always clear, it is simply masked, whereas simulation threatens the difference between the 'true' and the 'false,' the 'real' and the 'imaginary' (ibid.: 5).

It is the blurring of the boundary between the atmospheric real and the atmospheric imaginary that we see in digital emissions simulations. It is important to note here that digitisation and simulation are not merely coincidental events in British atmospheric government. As Baudrillard recognised, there is something about the invisible qualities and easy manipulation of the digital signal that make it particularly adept at facilitating the production of simulations.

Government always involves some form of simulation of reality in order to function. But it is important not to interpret this simulation as some form of State deception. In atmospheric terms simulation is all that we can know about the atmosphere at any significant systemic scale. To paraphrase Baudrillard, *the atmospheric simulacra does not hide the truth, it is the notion that there is an atmospheric truth that masks the fact that there is none* (ibid.: 1). If taken to its (il)logical philosophical extreme Baudrillard's notion of simulation can lead to a form of ontological nihilism within which the absence of atmospheric truth produces a debilitating relativism in air politics (see Soja, 2000). It is thus important to clearly establish at this point that I do not ascribe to the view that there is not an atmosphere with measurable ratios of emissions. Notwithstanding this point, I believe that the history of air government in Britain reveals that the ontological fabric of the atmosphere can never be known at the scales at which governmental knowledge regimes tend to be constructed. In governmental terms then, the removal of the atmospheric simulation does not unveil a governable air truth, but merely unmasks a debilitating *desert of the atmospheric real*.

Having specified more clearly how I understand the notion of atmospheric simulation, it is important to consider the computer-modelled atmospheric emulation within the formation of government rationalities. Luke (1995) provides one of the most detailed studies of the implications of computer-based simulation technologies on the arts and science of government. Through the notion of *telemetric territoriality*, he explicitly considers the impacts that bit-generated cyberspatialisations have on the practices and rationales of modern government. At the centre of Luke's analysis is a realisation of the profound implications of cyberspace for new forms of governmental being. He reflects,

> Software and hardware ensembles need to be reappraised not as inert combinations of machine instructions and instructed machines but rather as genetic operations, creating artificial environments for new social formations and digital beings out of their interactivities (ibid.: 55).

While cognisant of the profound implications of computer technologies for the operation of government, Luke focuses much of his attention on the State territorialisation of cyberspace. To this end he considers how notions of e-bureaucracies and wired republics reflect attempts to govern through the virtual environments of cyberspace (ibid.: 28). But what I am most interested in are the potentials of government-based cyberspatialisations for imagining new ways of governing the territorial fabric of the State itself.

Although simulated emission maps are produced, and continue to exist, within the cyberspatial *infostructures* of the NAEI, what is noteworthy is its ability to recast the territorial apprehension of the atmosphere by the British

State.[21] If we compare contemporary emissions maps with the rigid atmospheric regions of the NAPS, for example (see Chapter Six), it is possible to discern how cyberspatialisation can generate new territorially based images of atmospheric emissions. Gone are the fixed statistical registers of the NAPS, now replaced by the flexible vectors of pollution simulated by the NAEI. The NAEI's simulations emulate the dense concentrations of air pollution around the large conurbations of London, Birmingham and the North West; the purity of air on the North West coast of Scotland; the concentrations of pollution along major transport routes such as the M4 and M6 motorways, which act to spread pollution beyond metropolitan centres and into the British countryside. But this is not the territorialisation of telemetric technology it is the telemetric production of territory. As a telemetric production of territory, emissions maps not only embody the governmental colonisation of cyberspace, but also the cyberspatialisation of the governmental imagination of space. It is, in other words, a telemetric signal for new forms of governmental interventions within atmospheric space; a new basis for States' rationalisation of air government; a cyberspatial reason for government. Digital emission maps are essentially precursors to a more flexible geography of air government that, rather than acting on bounded territorial districts, seeks to intervene within air relations at the level of the metropolitan conurbation, the highway and the rural spaces subject to various pollution vectors.

As has been previously discussed, in his analysis of modern governmentalities Foucault described how government has increasingly sought to work with the socioeconomic realities within which it is situated, only intervening when the 'natural' course of events associated with neo-liberal being are artificially inhibited (2007 [2004]: 56–7). With the birth of computer-based technologies it appears that the time has come to not only consider the role of the real, but also the hyper-real within the practices and rationalities of government. According to Luke, such a step requires moving beyond the 'epistemic realism of the State' in order to reveal the power of simulation in the art of government (1995: 6). As Barry presciently observes, 'The production of scientific information does not mirror the world as it is, but forges something new, with more or less inventive consequences. It multiplies realities' (2001: 155). It is through the multiplication of new atmospheric realities that the NAEI has ushered in a new set of opportunities for imagining and implementing air government in Britain.

Changing Modes of Atmospheric Conduct in a Digital Age

Before completing this chapter I want to reflect briefly on the impacts that the automated production of digital emissions data has had on *fin de siècle* atmospheric modes of conduct. As was discussed in Chapter Four, in order to have governmental purchase it is necessary for new regimes of atmospheric knowledge to find modalities of influence through which to restructure the behaviour of those who use the atmosphere in different ways. Chapter Four illustrated how, during the late nineteenth and early twentieth centuries, clean air exhibitions were used as a way of informing urban citizens of the dangers of air pollution and how they could personally work to fight against the production of harmful atmospheric emissions. This section seeks to illustrate the new opportunities that the production of real-time atmospheric data, which can be immediately relayed throughout cyberspace, has generated for atmospheric conduct and subjectivity.

'Sensitive bodies', environmental decision making and the air pollution forecast

One of the most significant impacts that digital automation has brought to air pollution government in Britain has been its role in the production of emissions forecasts. As was observed in Chapters Three and Four, there has been a long historical connection between air pollution government and meteorology. While air pollution scientists have explored the connections between predicted atmospheric pressure systems and the severity of pollution events, the availability of real-time digital emissions data has transformed the science of pollution prediction. The availability of up-to-date air pollution data has enabled emissions figures to be synthesised with available weather system data in order to produce an ever more elaborate framework for forecasting daily levels of atmospheric pollution. There are two fundamental factors that determine the impacts of pollution on human health. First, there is the aggregate amount of different air pollutants that are produced in any given time frame. Second, the severity of pollution is also determined by the prevailing weather conditions, which dictate the speed with which pollution can move through the atmosphere and be dispersed (Defra, 2004: 7). It is in this context that Defra currently categorises high air pollution episodes into three types: *winter smogs*, *summer smogs* and *long-range pollution transport events* (ibid.: 7). Each of these pollution events is calibrated against specific weather conditions (see Table 7.2).

As Table 7.2 illustrates, weather is not just a contributory factor within high pollution events, but is actually the causal mechanism. In the context of winter smogs, high pressure weather systems serve to trap primary forms of air pollution in the areas that they have been produced within. In relation to summer smogs, the intense heat that higher pressure brings can activate the chemical reactions that produce secondary emissions such as ozone

Table 7.2   Air pollution types and weather conditions associated with high pollution episodes in the UK

Pollution episode type	Associated weather conditions	Typical forms of air pollution
Winter smog	Cold, still and foggy	Nitrogen dioxide, sulphur dioxide, particulates, volatile organic compounds (VOC) and other forms of pollution produced near ground level
Summer smog	Hot, sunny, still	Nitrogen dioxide, ozone, particulates
Long- and short-range pollution transport events	Hot and sunny with large-scale air movement	Nitrogen dioxide, sulphur dioxide, ozone, particulates

Source: Defra, 2004: 7

(ibid.: 7). It is in this context that the ability to calibrate real-time pollution levels with accurate meteorological predictions is so crucial to pollution forecasting.

At present the relative levels of severity associated with air pollution in the UK are classified on a 10-point scale (see Table 7.3). Three items are of particular interest within this system of atmospheric classification. First, the idea of banding the governmental assessment of air quality echoes Foucault's reflections on the construction of *bandwidths* of socioeconomic security (see Chapter Two) (Foucault 2007 [2004]: 6). As was discussed in Chapter Two, according to Foucault, modern forms of liberal government (or *apparatus of security*) are not dedicated to governing the *permitted* and the *prohibited* (an *apparatus of discipline*), but are instead involved in determining '[a] bandwidth of the acceptable that must not be exceeded' (ibid.: 6). Table 7.3 appears to reveal a threshold of atmospheric acceptability that stops with the classification of levels of atmospheric pollution at number 7 on this scale. The second feature of particular interest in this table, however, is how it differs from the paradigm of liberal government described by Foucault. Although the British government does have its own targets for acceptable levels for a cocktail of different air pollutants,[22] the air pollution forecast tends to shift governmental responsibility for atmospheric quality from the State to the *sensitive body*. The ability to be able to predict and classify the occurrence of *high* or *very high* pollution events moves air government from the realm of State bureaucracy and science to the level of individual air conduct. To put it another way, if the government can warn citizens of the likely crossing of atmospheric bandwidths of acceptability, it is as much the

Table 7.3   Air pollution bandings deployed by the UK for pollution forecast warnings

Band	Description
Low (1–3)	Minimal levels of air pollution that are unlikely to be detected even by those who are normally sensitive to emissions
Moderate (4–6)	Pollution levels may be noticed by *sensitive people*, but unlikely to need amelioration
High (7–9)	Significant impact on *sensitive people* who may be required to take ameliorative measures
Very high (10)	Worsening effects of emissions on *sensitive people*

Source: Defra, 2004: 4

responsibility of individual citizens to change their everyday atmospheric conduct (perhaps by walking rather than taking the car; going to the park for lunch instead of staying inside an air-conditioned office) in order to militate against the risk, as it is the State's to regulate air quality.

The third, and interrelated, point of interest associated with Table 7.3 is the very idea of the *sensitive body* it establishes. Aside from its rather pejorative undertones, the notion of the sensitive body again serves to shift atmospheric government from the State to the atmospheric subject. To an extent the assessment of likely effects of all forms of pollution is in part dependent on the individual subjected to emissions. Aside from the assiduous, long-term impacts of emissions on human health, it is clear that what counts as air pollution will be different for the asthma sufferer compared to the person with no respiratory health problems. In this context, it is necessary to pre-warn atmospherically vulnerable groups of the dangers of future air pollution events. But notwithstanding this point, it is clear that the notion of the sensitive body tends to relocate the source of the air pollution problem, and its potential solution, from the publicly regulated atmosphere and onto the respiratory-marginalised subject. In order to facilitate the shifting responsibility for atmospheric government that is enabled by the pollution forecast, the bandings revealed in Table 7.3 are published daily in national newspapers, on TV weather sections, on *broadcast teletext* facilities, and on the World Wide Web. This form of care against, rather than for, the atmosphere can be seen in a number of contexts ranging from the *sensitive body* of the asthma suffer, to campaigns to promote personalised forms of protection against the development of skin cancer. While these types of reflexive atmospheric risk management stem from the broad forms of government-sponsored medical care associated with biopolitics, they tend to operate at a much more personal scale of self-monitoring, calculation and decision making.

While the banding of air quality forecasts has proven relatively successful at predicting and governing air pollution events that are associated with winter and summer smogs, they have been less successful at foreseeing the consequences of the serious pollution events associated with the long-range transportation of emissions. A much-studied example of this occurred between 24 March and 2 April 2007 (see Defra, 2008). During this period, elevated levels of particulate pollution arrived in the UK, which was not anticipated by the AEA's team of air pollution forecasters. This pollution event caused either high or very high levels of air pollution in places as far apart as Manchester and Portsmouth, and Bristol and Glasgow (ibid.: 5). Air mass trajectory analysis provided by the Meteorological Office revealed that this pollution episode had emanated from agricultural fires in west Russia and the Ukraine, which had mixed with sandstorm particulates from north Africa and been brought to the UK by an easterly weather system (ibid.: 1). Although this pollution event was recorded by the British government's network of air monitoring stations, it was not forecast and no air pollution warnings were issued. One of the main reasons that this event was not forecast was that the AEA is dependent on NASA-supplied data on large-scale pollution events emanating from outside the EU. In this instance the NASA Natural Hazards website did not issue a warning, and when satellite images of the dust storm did appear it was felt that prevailing weather conditions in the UK would prevent a major pollution episode (ibid.: 1). This incident reveals the ever more complex spatial scales and technological relays through which the prediction of air pollution in the UK is now operating. While at one level, this story exposes the new pan-territorial scales at which British atmospheric government is active, it also indicates the dangers associated with placing too much faith in the reliability of the elaborate risk bureaucracies that are now emerging within modern State systems (see Gandy, 1999).

Real-time air data and the online citizen

In addition to the impact of air pollution forecasting, the production of publicly accessible real-time data on atmospheric emissions has also generated possibilities for new patterns of environmental decision making and conduct. The growth of automated air monitoring in Britain during the 1980s and 1990s corresponded with key international directives (particularly the Aarhus Convention of 1998) concerning the accessibility of government-produced atmospheric data.[23] The twin forces of automation and data democracy consequently ushered in a new era of public access to environmental knowledge in Britain. Suddenly, it was not just that environmental knowledge was being produced in real time, but that the public had a right to freely access the information that their taxes were ultimately paying to produce. The fact that the expansion of real-time air sampling during the 1990s corresponded with the rise of the World Wide Web also provided a new, rapid delivery mechanism for environmental data entering the public domain. While recent analysis suggests that forms of e-government tend to have only a relatively minor impact on the connections between the government and those who are being governed (*The Economist*, 2008), in this section I want to briefly consider what the Web-supported system of air knowledge could mean for atmospheric government and conduct in the twenty-first century.

It is 9.20am as I sit writing this chapter in the relatively clean air of Aberystwyth. I click on the Firefox icon at the base of my computer screen and search for the *UK Air Quality Archive*.[24] I am immediately presented with an interactive regional map of the UK. Highlighting the West Midlands region I find a map of the automated air sampling sites in the area. I select the automated sampler at Walsall/Willenhall (my home town) and ask to receive the latest hourly air monitoring data. This particular monitoring station is located on Johnson Road, about 700 metres from the ever-busy M6 motorway. Automated sampling has been conducted at this site on an unending loop since midnight, 24 September 1997.[25] In an instant I receive an atmospheric data set that was taken at 8am that provides me with precise information on the concentrations of various oxides of nitrogen in the air enveloping my birthplace. Before I have chance to record the figures taken at 8am they are replaced by those recorded for 9am, again instantly relayed to my computer screen (see Table 7.4). But what does this ability to move instantly in air space the 90 miles from Aberystwyth to Walsall to gain a real-time insight into the constitution of the atmosphere mean in governmental terms?

Table 7.4   Hourly averages air pollution data for Walsall (Willenhall), UK Air Quality Archive

Pollutant	Date	Time	Measurement	Unit
NO	04/03/08	9am	10	μgm^{-3}
NO_2	04/03/08	9am	34	μgm^{-3}
NO_x as NO_2	04/03/08	9am	50	μgm^{-3}

Note: This table has been adapted from http://www.airquality.co.uk/archive/result.php?site_name=Walsall+Willenhall&f_site_id=WAL2&f_area=last_hour&Submit=Submit (accessed 4 March 2008)

At one level the presence of open access, real-time air pollution data suggests the possibility of the emergence of a new of type of digital atmospheric self. In his discussion of the constitution of *digital beings* and subjectivities within the realms of cyberspace, Luke observes that digital subjects have been a feature of disciplinary government for some time, taking the form of statistical populations and persons that are constantly relayed through the data banks of government departments and ministries (1995: 35). Yet the digital atmospheric selves that are latent in the UK Air Quality Archive are a different type of being, not so much subjected to digital form, but potentially liberated by it. This is a digital atmospheric self that is able to overcome the frictions of atmospheric distance to compare the quality of air in its neighbourhood with that of other areas of Britain, Europe and perhaps even the wider world. It is possible to imagine this information being utilised to make key lifestyle choices concerning where to live, which school to send your children to, which park to walk in. In addition to providing atmospheric knowledge for lifestyle choices, it is also possible to imagine the Air Quality Archive being used as a basis for the mobilisation of new forms of air politics (see Chapter Nine). These air politics movements could use government data as a legitimate basis to protest against the unjust distribution of atmospheric pollution in certain neighbourhoods, or deploy the Air Quality Archive as a way of monitoring the impacts of new road, airport or factory developments. As a site for potential political action, the Air Quality Archive can serve to mobilise communities as agents of local atmospheric government themselves, using the tools of the State to monitor regular polluters, and also to ensure that the government is meeting its own commitment to air quality in different local areas. To these ends the atmospherically governed become the putative governors of State bureaucracies.

On this particular day, my own excursion into digital atmospheric surveillance does not prompt me to thrust myself into a struggle for air justice in my home town: the levels of nitrous oxides are deemed as low when I cross-reference them against the government's air pollution banding system. But I think care needs to be taken before we go too far into a celebration of the types of liberated and empowered atmospheric citizen that appears to be promised by these new digital developments. At a very simple level, our celebration should be curtailed through recognition of the geographical limitations of the automated network that supports the Air Quality Archive. Not only do automated sites not cover all areas, but not all sites measure the same range of pollutants. In classical Foucauldian terms, this is a network of surveillance that has been established to ensure atmospheric security, not the free availability of geographically complete atmospheric data (for more on the relationship between air pollution monitoring and community politics see Chapter Nine). In this context it is possible to see the Air Quality Archive as producing two types of atmospheric citizen: the digital, real-time citizen, who is plugged into subtle changes in the atmosphere; and the analogue citizen having to track down out-of-date data on local air quality in government reports. But, at another, perhaps more conceptual level, it is also important to critically question the types of atmospheric freedom and empowerment that can so easily be associated with phenomena such as the Air Quality Archive. Drawing on the work of both Foucault and Deleuze, Nikolas Rose analyses the power relations that continue to structure contemporary notions of freedom (1999a: 233–73). According to Rose, the notion and practices associated with freedom can in themselves reflect powerful modulations of contemporary governmental power and control. Consequently, whereas Foucault's *disciplinary society* has a clear demarcation between the government of conduct and spaces of discipline, Deleuze's vision of the *society of control* is based upon the control of conduct being inculcated within the individual through the everyday spaces of learning and training – the spaces in and through which we define our freedom (ibid.: 233). While this vision of everyday self-conduct and training is clearly evident in Foucault's later work on the ancient history *of self-care*, it is suggestive of new registers of power and government in the modern world. In this context, could the atmospheric freedoms associated with the Air Quality Archive actually be a suffocating form of atmospheric control within which the impacts of the air on all aspects of our life are to be constantly assessed and associated risks managed? If we accept this new interpretation of the atmospheric self, it is not just the position of the citizen that gets recast. Now government is less about the care of *each and all* in the atmosphere, and more about providing the routine air data, and calculating the tolerable bandwidths of existence, within which self-government is to be conducted.

Conclusions

Andrew Barry has argued that a *technological society* is a society that takes technological change as a prompt for new forms of political intervention (2001: 2). In this chapter I have sought to reveal how the rise of new automated and digital technologies for air pollution monitoring have altered the reasons for and mechanisms of air government in Britain. It is important to note that Barry does not deploy the notion of a technological society to denote a point of technologically enthused political change. As we have seen throughout this book, technologies – ranging from the Winkler tube to the jet dust counter; the standard deposit gauge to the smoke diagram – have all influenced the nature of air pollution government with science in Britain. In this context, this chapter does not seek to claim that the birth of digital automation facilitated a new prominence of the technological within air governance. Rather we have seen how these new atmospheric technologies have made it possible to govern the atmosphere and socio-atmospheric relations in new and dynamic ways.

We have also seen how the environmental revolution in British government ultimately resulted in the proliferation of automated digital sampling in the 1980s and 1990s. This system not only provided more atmospheric data, it also produced it in a form that could be easily utilised within computer-based systems of atmospheric simulation. The power of atmospheric simulation to model and predict the spatial distribution of air pollution means that the British State is increasingly governing the air through the construction of hyper-real, or telemetric, territorialities that combine measured and surrogate atmospheric statistics. In addition to facilitating the rise of simulated atmospheric governmentalities, this chapter has also revealed the potential impacts that real-time, online sources of air pollution data could have on the reconstitution of atmospheric self-conduct. At one level, the production of real-time air knowledge is enabling the creation of ever-more elaborate systems of pollution forecast that can be used to insulate sensitive bodies from the worst effects of severe pollution events. At another level, the ready availability of government-approved atmospheric data is enabling the emergence of atmospherically empowered citizens with the ability to use atmospheric knowledge as a basis for an expanded series of personal decisions. Within all of these processes of governmental change it is important to reconsider the role of the technological within new forms of political intervention. This chapter has clearly shown that atmospheric technologies are not simply inert tools, deployed to serve the preconceived strategies associated with changing forms of governmental reasons (see also Chapter Four). Automated samplers, computer processing units and the internet have all come together to actively reshape the nature and reasons for atmospheric government in the UK. Yet these technologies have required favourable socio-political circumstances to be deployed in governmental circles. So, from the publication of the *Limits to Growth Report* in 1972, and the London smog of 1991, to the passing of the Aarhus Convention in 1998, novel technologies have not simply recast atmospheric government, but are themselves legitimated and animated by political circumstance. Consequently, and as the final section of this chapter stressed, while it is easy to celebrate the liberating potential of automated and digital technologies of British air government, the impact of new technologies is always contingent upon circumstance, and in need of careful examination. It is for this reason that axiomatic recourse to either technophobia or technojoy needs to give way to careful analyses of both the governmental reasons for deploying technologies and the governmental affects that technologies produce.

Chapter Eight

Environmental Governmentalities and the Ecological Coding of the British Atmosphere

Introduction: Ecology in an Age of Governmental Revolution

If you happen to be travelling south on the A453 between Nottingham and East Midlands Airport and take a brief detour to the village of Sutton Bonnington you will come across two rather peculiar looking fields. While the arable crops being grown in these fields (perhaps at the time sugar beet, wheat or oats) may be in perfect keeping with the agricultural landscape of the English East Midlands, it is what is located within the two fields that will undoubtedly appear most incongruous. At the heart of the crops and where the two fields intersect is a Portakabin positioned next to a 3-metre-high instrument mast with a bewildering array of clamps, tubes and devices attached to and surrounding it. This complex instrumental armature of gas analysers, spinning wind-measurement devices, radiation detectors and thermometers provides one of a series of field sites in which university scientists, working as part of the *Centre for Ecology and Hydrology*, have been developing new systems of atmospheric monitoring and scientific analysis in the UK. The Centre for Ecology and Hydrology (CEH) is a collaborative research centre, supported by the government's Natural Environmental Research Council (NERC), which brings together scientists from different universities in order to construct large-scale studies of various forms of environmental change. As its name suggests, the CEH is emblematic of a new relationship between atmospheric science and government in Britain: a relationship that is grounded upon the principles and allied analytical techniques of ecology. This chapter explores the extent to which the ecological sciences being practised on a day-to-day basis in fields like those found in Sutton Bonnington, embody the foundation of a new regime and associated rationality of atmospheric governmentality in Britain.

Donald Worster famously observed that ecology is a 'peculiar field' of scientific and para-scientific study (1994: ix). Indeed, the various forms and varied histories of ecology make analysis of its role within the formation of new frameworks of atmospheric intelligibility in Britain problematic. Although identifying the arrival of the term ecology in 1866, Worster asserts that the *idea of ecology* can actually be traced through a range of early modern philosophical and scientific traditions that he refers to as the 'penumbra of ecological thought' (ibid.: x–xi). Given the broad sweep of thought that has become bound up with purportedly ecological ways of perceiving the environment, providing a succinct and adequate definition is difficult. Deliberately keeping delimitations of ecology broad at the moment, it is useful to approach it as a scientific and intellectual tendency that,

> [e]merged as a more comprehensive way of looking at the Earth's fabric of life: a point of view that sought to describe all of the living organisms of the Earth as an interacting whole, often referred to as the 'economy of nature' (ibid.: x).

Although it is tempting to assume that such a holistic environmental science would be connected to the emergence of a more ecologically caring set of governmental structures and practices, Worster asserts that the rise of the science of ecology within modern State apparatus is actually connected to a much more pernicious set of activities. According to Worster the *age of ecology* in government started in the 1950s when, in the wake of the Manhattan Project, a series of States started to commission ecological research into the environmental impacts of nuclear fallout and the movement of radioactive material through the various webs of nature (ibid.: 342–6, for a broader discussion of the relationship between ecological thought and military science see Seager,[1] 1993). The ecological study of the effects of nuclear fallout in countries like the USA and Britain in the 1950s had, of course, direct links with emerging forms of atmospheric government. The study of the air-borne spread of Strontium-90, for example (produced by atmospheric nuclear tests), became an important part of the air monitoring apparatus of Britain and the USA in the 1950s and 1960s.

If Worster is correct, then it appears that the emergence of ecological science within, at least Western, governmental apparatus significantly predates the ported environmental revolution in British government of the late 1960s and early 1970s discussed in the previous chapter. While the presence of government-sponsored ecological science may have existed in Britain during the 1950s, its does seem that key scientific and political events during the 1960s had a significant impact on the links between government and ecology in Britain. At one level it is clear that the significant rise of green political groups and grassroots environmental movements during the second half of the 1960s placed renewed political pressure on the State to rethink the ways it monitored the environment. But more specifically in relation to ecology, it appears that it was the systems of knowledge production upon which these environmental groups were based that would have the most lasting impact on the environmental knowledge-gathering apparatus of the British government. While governmental ecology may have been a product of the nuclear age, the government was presented with a new breed of ecological science during the mid to late 1960s. This was a strain of ecology that was grounded in the work of civic sciences such as Rachel Carson, and which did not use ecology as a framework to expose the unusual effects of the alien technologies of nuclear technology, but instead the insidious effects of the chemical byproducts that had become an accepted aspect of industrial society (see Carson, 2000). It is in this context that the work of ecological scientists working beyond the institutional remit of States appeared to simultaneously alert governments to the environmental crises of industrial society and to provide the scientific paradigm within which it could be studied and addressed. This chapter asserts that if the environmental revolution in British atmospheric government embodies more than just a superficial set of institutional reformulations and ideologically pronouncements, it is a revolution that must have evidence of ecological practices and rationalities at its heart.

Focusing on the case of British air pollution science and government, this chapter explores the extent to which it is possible to discern the emergence of a new ecological rationality in Britain. In keeping with the broader ethos of the book, however, analysis presented here seeks to avoid reading off a new ecological governmentality merely from institutional shifts in the form of government, and instead focuses on the sciences that fuse government and knowledge production as the basis for interpreting changes in governmental reason. In focusing on this scientific nexus, this chapter considers the extent to which ecologically inspired changes in the practices of governmental knowledge production either reflect or facilitate changes in the reasons for air pollution government. In this context, it is as much an analysis of whether it is possible to discern the operation of an ecological governmentality in British air pollution policy as it is about correlating contemporary writings of green governmentality with a British case study.

The chapter begins by outlining the interrelated concepts of environmental governmentality and ecopolitics. The notions of environmental governmentality and ecopolitics (or *ecopower*) have emerged from a body of work that has sought to apply Foucauldian theories of power and government to various environmental questions (see Rutherford, 1999). Such concepts provide frameworks to explore and analyse the changing environmental parameters of atmospheric governmentality in Britain. Analysis then moves on to consider the institutional contexts within which it is claimed a new ecologically imbued rationality of government action towards the atmosphere has emerged. As with the previous chapter, this section will consider the impacts of the CUEP, DoE and RCEP, but it focuses much more on the impacts of these institutional transitions on governmental conceptions of the atmosphere as opposed to air monitoring technologies. The following section considers the extent to which it is possible to discern a shift in ecological rationality within British air government by exploring recent changes in the focus and practice of government-sponsored air pollution studies. Drawing on the example of new ecologically based programmes for air pollution monitoring utilised by Defra, NERC and AEA Energy and Environment, this section considers the extent to which these changing practices of air surveillance constitute the basis for a revised rationality of air government.

Ecopower and the Atmosphere as a Site for Ecological Government

Much has already been written on the idea of ecology as a basis for new forms of *fin de siécle* governmentalities (Agrawal, 2005; Darier, 1999; Rutherford, 2007). At the heart of these analyses is the belief that the environment constitutes a new arena of care and calculation within governmental activities, and that this nexus of care and calculation can be discerned in a new era of environmental ministries and ecological specialists employed by government. Yet despite the contemporary prescience of this intellectual project it has been characterised by some confusion and intellectual imprecision. For example, we routinely find reference to notions of green governmentality, environmental governmentality, ecological governmentality and even the neologism environmentality being used interchangeably. Whether these variously prefixed mentalities are supposed to refer to the same or different sets of governmental processes and rationalities is, however, unclear. Interspersed within this new lexicon of government are the equally interchangeable notions of ecopower and ecopolitics. In part, I believe that the confusion surrounding the attendant terms deployed within this nascent school of Foucauldian enquiry has actively contributed to uncertainties over the precise relationship between ecopower and biopower. Building on some of the groundwork provided by Chapter Two, this section attempts to clarify what the notion of ecological governmentality, as a distinctive Foucauldian concept, may actually mean and how it relates to interconnected sets of Foucauldian concepts.

Rather than focusing on its green, ecological or environmental prefix, I believe that the key to beginning to unpack the notion of ecological governmentality is to return to the complex definition of governmentality provided by Foucault (see Chapter Two). Although Foucault equates the phrase governmentality with an actual expression of governmental form or, 'the ensemble formed by institutions, procedures, analyses and reflections, calculations, and tactics that allow the exercise of this very specific, albeit very complex, power [...]' (2007 [2004]: 108), it is inappropriate to use this paradigm of governmentality as a pre-given basis for exploring the regulation of new governmental objects such as the environment. The reasons why this strategy is inappropriate are threefold. First, Foucault explicitly states that his excavation of governmentality was an attempt to uncover a system of power and knowledge that has population as its target. Notwithstanding this point, I do not want to suggest that it is necessary to conduct a new history of government so we can understand the emergence of a mechanism of power and knowledge that takes ecological processes as its target. After all, it is clear that the same era of government that was able to envision a construction of population as a horizon for government was part of the same historical process that governmentalised the environment. Second, in addition to stressing the notion of governmentality as a condition of political existence, Foucault also emphasised to his audience that governmentality is a part of a process and tendency within political history. As an ongoing process and tendency it would seem important to recognise the constant changes that are reshaping the governmental condition. My question would thus be: what changes in liberal and neo-liberal government have occurred since the emergence of biopolitics and what impacts have these changes had on the emergence of the environment as a new horizon of governmental concern? Thirdly, and most importantly, to use governmentality as a static paradigm for comparative study of ecological government undermines the whole ethos of Foucault's governmental history. As was outlined in detail in Chapter Two, Foucault's 1978 (and 1979) lecture series were not primarily concerned with delimiting the condition of liberal governmental power, but in explicating a method for the study of government history (ibid.: 358).

If governmentality is interpreted as an historical method as much as a political paradigm it is clear that it is not enough to simply assert that an ecological governmentality emerges in Western States such as Britain during the late 1960s and early 1970s. To do so would be to subscribe to what Foucault describes as a discourse of history, rather than excavating a history of discursive practices.[2] To put it another way, to take the fact that the environment starts to be discussed in different ways in Britain in the late 1960s, and begins to influence the institutional structures of government, as evidence that a moment of transformation (or 'revolution') has arrived (before which the environment is absence from State discourse and practices, and after which it is a powerful force within government) is to short-circuit Foucault's own historical method. The notion of ecological governmentality should thus not be used to organise histories of both the environment and government, but should instead act as a stimulus to the study of the historical processes and mechanisms of power that generated a will to *ecological knowledge*. By studying the history of the government of the atmosphere as a socio-environmental system, this volume has already provided a history of the practices of science and government that informed the revolution in air pollution government in Britain. In developing this history it has been possible to see that the *governmentalisation of the environment* – understood as the subjugation of environmental systems to rigorous scientific codification and classification – has a long history in Britain. The question which remains to be explored is whether, and if so why, has there been a steady ecologisation of governmental reason in Britain over the last 40 years?

Having established how the notion of ecological governmentality is approached in this chapter, it is now important to be clear about how the allied concept of ecopolitics is interpreted. While it is tempting to suggest that ecopolitics is to ecological governmentality what biopolitics is to governmentality – namely the political mechanisms and diffuse networks of power through which particular forms of governmental reason are expressed and realised (see Huxley, 2006) – this characterisation would represent a gross simplification. It has already been argued by numerous scholars that rather than superseding biopolitics, ecopolitics should be understood as a deeply interconnected corollary to its historical predecessor (Rutherford, 1999). It is in this context that Darier observes that,

> Current environmental concerns could be seen as an extension of 'biopolitics', broadened to all life forms and called 'ecopolitics'. On this scenario, the normalizing strategy of ecopolitics is the most recent attempt to extend control ('management') to the entire planet (Darier, 1999: 23).

Reflecting on the related work of Rutherford, Darier goes on to suggest that such governmentalities involve,

> Building on the Foucauldian concept of 'biopolitics', but pushing beyond its central concern with human life [...] the current interest in ecology can be characterized as an 'ecological governmentality' in which all life forms become objects of scientific enquiry, a series of state calculations based on 'security' and on the disciplining/normalization of the population (ibid.: 28).

Three important points for discussion arise from Darier's characterisation of ecopolitics. First, is the notion that ecopolitics builds on the Foucauldian concept of biopolitics. This conceptual building process involves utilising Foucauldian theories of power (relating to the scientific production of knowledge, diffuse networks of control and systems of self-conduct), but applying such ideas to issues of environmental as well as biological management (see, Rutherford, 2007: 297). The second, and arguably most problematic (see below), implication of Darier's reflections is that eco-power tends to operate at new, increasingly global scales. Third, and inter-related to questions of scale, Darier suggests that ecopower involves the *pushing beyond*, or extension of, biopolitical knowledge-gathering and governmental apparatus.

The idea that ecopolitics involves an extension of biopolitics – or government *pushing beyond* the spaces of existence associated only with human life – requires further critical reflection. The connection between ecopolitics and an extended life politics is important because it provides the foundation for ecological governmentality as a concept. But what does this politics of extension actually entail? First, although there is an obvious temptation to perceive of ecopolitics as a form of spatially extended – or upscaled – biopolitics, it would be inaccurate to assume that ecopolitics necessarily operates at a higher spatial scale to biopolitical strategies. The rescaling of governmental concerns to supranational levels is not a unique feature of ecological governmentalities. Despite its strong association with the management of national populations, for example, is it not also possible to perceive of elements of biopolitics operating at more global scales in relation to eugenics, the work of the World Health Organization, and even the contemporary struggle against HIV-AIDS? What differentiates ecopolitics and biopolitics is thus not the scales at which they operate, but the scope of the environmental considerations they embrace. In this context, it is equally valid to explore the differences between bio- and ecopolitics at micro-scales. It is for example possible to imagine both bio- and ecopolitical concerns operating at the micro-biological level in relation to the production of genetically modified (GM) crops and plants: with biopolitical rationalities focusing on the human health impacts of GM food and ecopolitical mentalities considering the impact of GM monocultures on the long-term diversity and sustainability of entire ecosystems (note that this concern with ecosystems does not involve the abandonment of the biopolitical, but the recalibration of human environments in relation to the needs of broader environmental systems). The governmentalities associated with notions of ecopolitics are thus always about a recalibrated, and ecologically complex, reason for government, and never purely about an extended spatial remit of governmental power.

In discussions of ecopolitics it is important to consider the practical impact that such reasons for government have had on the way governmental knowledge about the environment is gathered. In his analysis of the relationships between bio- and ecopolitics, Rutherford (1999) recognises that the environmental revolutions that swept through governments such as Britain's in the early 1970s, were not the first appearance of environmental discourses in governmental strategy. According to Rutherford, the rise of modern biology in the eighteenth and nineteenth centuries saw increasing recognition given to the fact that '[o]rganisms are functionally linked to their external surroundings' so as to 'exchange resources with their environment' (ibid.: 40). Rutherford also describes how concerns over available environmental resources expressed during the nineteenth century embodied a much earlier environmental revolution in European governments (ibid.: 51–60). What differentiates these environmental thought-waves from those that emerged in the twentieth century was that they were connected (often through Malthusian ideologies) directly to questions of the fecundity of national population. The notion of environmental resource, for example, was essentially a product of biopolitical calculation. The emergence of questions of environmental resource availability within rapidly industrialising States led to the formation of new governmental modes of environmental calculation which were connected to the sciences of geology, pedology and mineralogy. It is precisely for these reasons that I choose to use the term ecological governmentality as opposed to environmentality. If concern with the environment is not a new reason for government then any claim to revolution in governmental rationality in Britain during the late 1960s must be marked by the growing influences of the sciences of ecology. While these new frameworks of scientific intelligibility targeted ever more diverse aspects of the natural environment, they were characterised primarily by a new rationality of how the environment was to be studied, assessed and governed. Rutherford goes on to claim that scientific ecology offers the *intellectual machinery* required for modern environmental governmentalities to exist (ibid.: 37). As a holistic science dedicated to the study of the complex interactions and symbioses of biological and environmental systems, if there is a new environmental governmentality associated with the British atmosphere then ecology appears to offer the most likely context for a revised epistemic system of air study.

In the remainder of this chapter I want to explore the idea of ecological governmentality in Britain by tracing the new types of scientific practices and knowledge-gathering apparatus upon which it may be based. Timothy Luke provides an indication of the types of knowledge that are symptomatic of an ecological governmentality when he describes contemporary attempts by State administrations to eco-code the knowledges they routinely produce (1999: 134). The idea of eco-knowledge, or ecologically coded knowledge, provides a key organisational framework for the remainder of this chapter. The idea of eco-knowledge is important for two reasons. First because it places immediate emphasis on the role of ecological science in knowledge-gathering regimes associated with ecopower and governmentality. Second, in addition to denoting a new scientific structure within atmospheric government, the idea of the eco-coding of knowledge also suggests that air pollution studies should be marked by new systems of atmospheric monitoring that operate through a series of ecological filters and measurement grids.

Holistic Sciences and Integrated Government: On the Institutionalisation of Ecological Thought

The national scale of revolution

While this chapter deliberately seeks to avoid reading off changes in the ecological governmentality of the British State from changes in its institutional fabric, it is nevertheless important to consider whether the emergence of new bureaucratic structures in government either reflected or facilitated the emergence of a more ecologically coded governmental system. Chapter Seven briefly discussed the formation of a new environmental armature within the British State that centred in the Department of the Environment, but also included the Central Unit on Environmental Pollution and the Royal Commission on Environmental Pollution. It was the manifestation of such novel environmental bureaucracies that provided the basis for the popular description of an environmental revolution in British government. As we have already established, however, if these structural changes in the form of the State are synonymous with a *revolution* in environmental government – as opposed simply to a more direct focus on an already existing set of socio-environment concerns – it is to be expected that they will both reflect and support an ecological *gestalt*.

The institutional realignment of government associated with the environmental revolution was an initiative led by the Labour administration of Harold Wilson. According to Owens and Rayner, at the centre of this revolution was a concern that the structure of government – particularly around specialist departmental remits – made it difficult to govern the complexities associated with interrelated environmental systems (1999: 7). While it is difficult, if not impossible, to determine whether this attempt to redesign government was inspired by the sciences of ecology, which were being increasingly popularised at the time, it is at least possible to trace the official justifications that were made of these bureaucratic changes and to consider the relationships between such justifications and ecological thought. The first significant moment in the environmental revolution in British government came in 1969 with the establishment of the Central Unit on Environmental Pollution (CUEP). The CUEP inherited diverse, and highly complex, environmental pollution monitoring networks spanning aspects of fresh and marine water, terrestrial ecosystems and air pollution. With many of these networks adopting different pollution measurement standards, and reporting to different departmental committees and sub-committees, it is important to note that one of the first tasks of the CUEP was to gain a clear picture of how environmental monitoring in the UK actually worked. In the CUEP's 1974 report, *The Monitoring of the Environment in the UK*, the Unit stated,

[...] although there are many monitoring programmes in existence, these have been designed with little regard for how, taken together, they can be used to provide a coherent picture of national trends. We are therefore proposing to strengthen the machinery for coordinating the programmes in different sectors to ensure that the data collected is in a compatible form [...] (Department of the Environment, 1974: iii).

The evident desire of the CUEP to integrate monitoring systems that generated separate pollution data streams for the atmosphere, hydrosphere and biosphere could be interpreted as part of a gradual shift in the mentalities of environmental government in Britain. As the fledgling British environmental movement, supported by various branches of civic science, began to show the interconnected ecologies of industrial pollution, it became increasingly apparent that environmental government could not be conducted on artificially isolated systems such as the atmosphere. To this end, the review, and attempted integration, of environmental pollution monitoring undertaken by the CUEP would appear to reflect, intentionally or not, a fledgling ecological armature of the State. But while the CEUP may represent, at an institutional level at least, the partial ecologisation of British environmental monitoring systems (or the re-governmentalisation of the environment), it remains doubtful that it constituted a new reason for environmental government (the ecologisation of government). Returning to the relation between ecopower, biopower and the sciences of ecological government outlined above, it appears that the CUEP's desire for a more integrated monitoring system was still rooted in a strong biopolitical rationale. Accordingly, in 1974 the CUEP stated,

[t]he direct effects of pollution on human health, although important, are not the only things with which we must be concerned. Man is sustained by other species in his environment – by crop plants, domestic animals and fish, by a myriad of species often small and inconspicuous, that renew the oxygen in air and water, break down dead matter, recycle essential elements and are at the base of the food chains leading to ourselves and our livestock. If there is significant ecological disruption on a wide enough scale through pollution, man himself cannot escape hazard (Department of the Environment, 1974: 6).

Within the masculine appropriation of 'his environment' suggested within this passage, we see that the initial emergence of a purportedly ecological rationality of government in Britain was not about the abandoning of biopolitics in favour of an extended ecological realm of moral concern. At least initially, it appears that the rise of environmental governmentality in Britain was based upon an enlarged field of biopolitical calculation within which the government of the human environment was extended from the immediate health effects of pollution to embrace the broader network of ecological processes that make human life on this planet possible. To this end the beginning of Britain's environmental revolution appears not to have been predicated upon a new ecological reason for government, but the co-joining of an ecological sensibility with a biopolitical governmentality.

In 1970, the year after the formation of the CUEP, we find another watershed in the history of environmental government in Britain: the establishment of the Department of the Environment (DoE). The DoE was not only the first fully fledged ministry dedicated to environmental issues in Britain, but also in the world. Yet even in this, apparently defining, moment of environmental government, it is possible to discern a disjuncture between the changing institutional form of government and the prevailing reasons for governmental action. The DoE was formed by the incoming Conservative administration of Edward Heath and at a superficial level, at least, appeared to signal the ecologisation of government commenced by the previous Labour administration. The DoE brought together the environmental remits of 10 existing departments in order to provide a more coherent institutional context for environmental policy development and delivery. According to McCormick (1991), however, the formation of the DoE reflected the Heath administration's desire to consolidate central governmental control over myriad agencies, through the formation of super-ministries, more than any real commitment to a new era of ecological government. McCormick reflects, 'Despite the name, the creation of the DoE was more a reorganisation of government machinery than the creation of a new department with new powers. Many key environmental concerns were left with other departments' (ibid.: 16). Furthermore, Jordan (2000) points out that although the principle of an extended ecological sphere of government was seen by many as the rational basis for this new department, many subsequent Ministers of the Environment tended to focus much more on the connections between health and the human environment to the detriment of a more holistic governmental outlook. It is interesting to note that throughout the history of the DoE this basic tension has continued as it has developed a bifurcated focus on urban and regional policy on the one hand and ecological concerns on the other. It appears that the clarity of ecological thinking, which a more holistic form of environmental government was initially intended to provide, has been subsumed within the opacity that comes from a messy and confusing ministerial portfolio.

Perhaps the clearest sign of the emergence of a new ecological rationality in British government can be discerned within the third key institution associated with this revolution, the Royal Commission on Environmental Pollution (RCEP). As outlined in the previous chapter, the RCEP's role is, 'To advise on matters, both national and international, concerning the pollution of the environment [...]'.[3] Although, as we have established previously, the international scope of the RCEP's understanding of the environment should not necessarily be associated with a clear change of governmental rationality, it does appear to belie a certain degree of commitment to a more integrated and holistic way of understanding the operation of environmental systems that we could reasonably associate with ecological science. The RCEP's commitment to ecologically based interpretations of environmental destruction is, in part, supported by the definition of pollution it employed from an early stage. The RCEP sought to position its understanding of environmental pollution in relatively broad ecological terms stating,

> The Commission has interpreted 'pollution' broadly as covering any introduction by man into the environment of substances or energy liable to cause hazards to human health, harm to living resources and ecological systems, damage to structures or amenity, or interference with legitimate uses of the environment.[4]

Through its various reports the RCEP has used this definition of pollution to conduct research on various aspects of the human and non-human environment. While the RCEP occupies something of an ambiguous position on the fringes of government, the research of Owens and Rayner has revealed that the Commission has been responsible for developing a distinctively eco-modernist mindset within British government towards environmental policy and pollution control (1999, 11–13). While at one level this eco-modernist mindset asserts the economic value of astute environmental policy, it also suggests a much more holistic view of the governance of environmental systems.[5] This sense of holistic policy thinking is typically associated with the active governmental encouragement of industrial polluters to think through the diverse ecological vectors and socio-environmental feedback loops which ultimately mean that supposedly uncosted pollution carries heavy financial burdens for polluters – costs which are often carried into the future.

The institutional reformulation of environmental government in Britain ultimately led to calls for the formation of more ecologically integrated systems of air, marine and terrestrial pollution measurement and analysis. In its 1974 report, for example, the CUEP recommended the formation of a new environmental knowledge management system in Britain (Department of the Environment, 1974). Partly modelled on the National Air Pollution Survey, this new system was to be composed of a *National Environmental Focal Point* (eventually named the *National Data Network on Environmentally Significant Chemicals*); *Data Management Groups* for air, fresh water, marine and land pollution; and *Lead Laboratories* to coordinate data collection in different environmental sectors (ibid.). It was anticipated that the scientific and administrative harmonisation of environmental data collection envisaged by the CUEP would enable Britain to fit more effectively into the new supranational environmental monitoring networks of the 1970s (see below), as well as develop a more ecologically integrated view of the interconnections between pollution in different forms of environmental systems.

Following on from the CUEP's report, the RCEP's *Fifth Report* (*Air Pollution Control: An Integrated Approach*) specifically addressed the need for greater integration and standardisation in the collection of atmospheric pollution data (Royal Commission on Environmental Pollution, 1976). Closer inspection of the 1976 report does not suggest that it is a piece of pure ecological rationality. Strong emphasis is placed, for example, on the relationship between atmospheric pollution and economic development, human health and social amenity. Even the RCEP's call for greater monitoring integration appears to derive as much from a concern that the abatement of air pollution could see contaminates displaced to other environmental sinks as from a desire to develop an ecological account of the impacts of pollution (Royal Commission on Environmental Pollution, 1976: 3). However, the emphasis placed by the report on the impacts of air pollution on plant and animal life, and the calls that it made for greater synthesis in environmental knowledge production, mean that it did represent an important moment in the history of atmospheric pollution government. It appears that in moving beyond its remit on air pollution monitoring in 1976 the RCEP was beginning to lay the political grounds for new types of environmental monitoring science in government.

The recommendations of both the CUEP and RCEP on the need for more integrated systems of inter-media pollution monitoring in Britain did not call for a change in the sciences of pollution surveillance, but rather for the restructuring of the institutional structures through which environmental knowledge was being produced, transferred and compared. Notwithstanding this, I would assert that the evident desire for greater integration within pollution monitoring networks in Britain did reflect perhaps the most obvious indication of the rise of ecological rationalities within British government at the time. It is important, however, to note that the key recommendations made by both the CUEP and the RCEP – particularly the suggestion that the work of local authorities and *Her Majesty's Alkali and Clean Air Inspectorate* be amalgamated – were largely ignored by the British State until 1987 when *Her Majesty's Inspectorate of Pollution* was formed (Owens & Raynor, 1999: 14). Even at this point, it appears that the integration of environmental monitoring science in Britain was largely driven by the vision of a more efficient and cost-effective bureaucratic system (ibid.). To this end, it appears important to interpret the environmental revolution in British government to be at best a delayed revolution. This delayed revolution may have been made possible, and even glimpsed at, within the sweep of institutional changes that occurred in British environmental government in the early 1970s, but it does not appear to have been supported by a significant change in either the institutional context or working practices of pollution science for at least a decade.

Transnational science the international revolution

An analysis of the official discourses of the key institutions involved in the purported environmental revolution in British government reveals at best an uneven sense of change in the ecological rationalities of British government during the late 1960s and 1970s. In order to appreciate fully the impacts of ecological science and research on British atmospheric government in the 1960s and 1970s, however, it is crucial to move beyond the national institutions so far discussed. As Chapter Seven briefly outlined, by the early 1970s Britain found itself within an emerging space of international exchange and cooperation concerning atmospheric policy and air government. While it is important to recognise the impact that key events such as the *United Nations Conference on the Human Environment* and Britain's ascent to European Community membership had on atmospheric policy (see Jordan, 2000), it is also crucial to consider the impacts that these new international spaces of dialogue had on transnational scientific exchange. There are essentially three key international contexts of atmospheric government and scientific exchange within which the British State started to operate during the 1970s: the overarching global bureaucracy of the United Nations Environment Programme (which emerged from the 1972 Stockholm Conference) and the Organisation for Economic Coordination and Development (OECD); the European Economic Community; and the United Nations Economic Commission for Europe.

At a global scale the formation of the OEDC's *Environment Committee* in 1970, and the United Nations Environment Programme (UNEP) in 1974, led to calls for the establishment of a Global Environmental Monitoring System and International Register of Potentially Toxic Chemicals (IRPTC). At a European level, 1972 saw the formation of the *European Community Environment Programme* (ECEP), and the establishment of the *European Chemical Data Information Network* (ECDIN). The formation of the ECEP would provide the context for the long-term rescaling and recalibration of atmospheric monitoring science in the UK. In addition to informing the geographical spread of new standard techniques of air surveillance, the ECEP also constituted a key sphere for the adoption and extension of ecological techniques.

Operating at a scale between the European Community and more global collaborations, the United Nations Economic Commission for Europe (UNECE) also provided a crucial institutional context for the recalibration of British air monitoring. The UNECE was established in 1947 and is constituted by 56 Member States from the contemporary European Union, North America and Eastern Europe (United Nations Economic Commission for Europe, 2008).[6] While initially focusing on issues of economic cooperation and development, during the 1970s the UNECE started to instigated new forms of scientific and governmental collaboration concerning environmental security and pollution control. Crucially, the UNECE's evolving environmental remit was influenced by the findings of newly emerging ecological sciences in Europe. Of particular significance in this context were scientific studies conducted during the 1960s which revealed that emissions of sulphur produced within European industrial centres were travelling through the atmosphere for hundreds of miles before they eventually contaminated the fragile ecologies of Scandinavian lakes (ibid.). Concerns over the impacts and regulation of transnational air pollution consequently became a central concern at the 1975 *Conference on Security and Cooperation in Europe*, which was convened in Helsinki. While this conference is often discussed as a key moment of cold war cooperation and rapprochement, environmental concerns and ecological principles were a prominent part of the discussions. Section 5 of the *Final Act* of the Helsinki Conference – specifically covering environmental issues – emphasised the importance of developing interdisciplinary approaches to the study of environmental pollution; the production of internationally comparable data on pollution; and the importance of preserving key ecosystems and biospheres (United Nations Economic Commission for Europe, 1975: 29). While both the interdisciplinary ethos and data-sharing capacities of the Helsinki Agreement clearly reflect the impacts of ecological thought on international discussion of pollution abatement, the emphasis that the meeting placed on the notion of the biosphere is also instructive. This clearly reflects the influence of Soviet ecological science on discussions concerning pan-European pollution. In many ways the biosphere constituted the Soviet scientific equivalent of the ecosystem concept in the West, and while still framed within the holistic sciences of ecology, it placed particular attention on the role of geology within the long-term constitution of life and environmental sustainability.[7] What is most significant about the presence of the biosphere concept at the Helsinki conference, however, is that it reveals the crucial role of international organisations such as the UNECE in the cross-fertilisation and promotion of different ecological techniques and frameworks during the 1970s.

Perhaps the clearest example of the relationship between the UNECE, atmospheric government and ecological science can be discerned in the 1979 Geneva Convention on Long-Range Transboundary Air Pollution (LRTAP). LRTAP sought to construct a pan-European/Atlantic framework for monitoring and regulating transboundary pollution. The 1979

Convention built on the agreements forged in Helsinki, but resulted in ecological considerations becoming more central to systems of atmospheric science and government. The ecological dynamics of the 1979 Convention, and its subsequent Working Groups, can be discerned in two main ways. First, it stressed the importance of understanding the role of ecosystems as receptors of air pollution, and of analysing the direct ecological impacts of such pollution. Second, it stressed the value of measuring air pollution alongside pertinent meteorological, physico-chemical and biological data, so that a more complex picture of the variable impacts of atmospheric pollution could be developed.

Given that Britain was a member of the European Community, and signatory at the 1972 and 1975 UN Conferences and the 1979 LRTAP, its national air pollution science was critically shaped by such international initiatives. At the simplest of levels, the impact of such frameworks on British air pollution science can be discerned in the renewed efforts that were made to synchronise British atmospheric data with international monitoring regimes and to mimic the integrated systems of air monitoring they were promoting. It is important to acknowledge though that such changes in the ecological fabric of British air government were only continuations of the systems of institutional restructuring (around the CUEP, DoE and RCEP) that were already under way in the UK. As we will see in the following section, however, it was Britain's integration within new, international systems of scientific exchange and cooperation that would ultimately lead to some of the most profound transformations in the governmental sciences of air pollution.

Ecological Science and the Changing Techniques of Atmospheric Government in the UK

There are two main reasons for doubting the existence of an ecological revolution in the rationalities of air government in Britain during the 1970s. First, it is far from clear that the institutional changes in the apparatus of environmental government were under-girded by anything but a notional reference to ecological science. Second, the more ecologically imbued ethos of European scientific cooperation evident in the mid-1970s took a significant amount of time to filter into the working procedures of national scientific practice. As with the emergence of automated and digital air monitoring charted in the previous chapter, we have to wait until the mid-1980s before it is possible to detect a real shift in the techniques deployed by government towards air pollution science. In this section I claim that the rise of techniques of ecological science can be discerned in British air government in three ways: (i) in relation to the use of new, environmental locations as the basis for air pollution monitoring; (ii) in relation to the assessment of the atmospheric thresholds at which certain ecosystem types start to suffer and collapse; and (iii) in relation to the greater methodological sensitivity shown to the influence of various ecological factors on the socio-environmental impact of atmospheric emissions. The remainder of this section explores each of these traces of ecological science in turn.

Relocating air surveillance: ecology, ersatz and a new monitoring environment

The history of British atmospheric government traced in this volume has been characterised by two dominant trends. First, there has been a geographical focus on the scientific assessment of the quality of metropolitan air. From the earliest times of smoke inspectors in Victorian Britain, it was cities that occupied the spatial hub of governmental concern with the quality of the air. As both the main sources of industrial air pollution and home to the populations to which governmental care was being extended, it was inevitable that cities should be at the geographical centre of air pollution government with science. Second, there has been an inexorable movement of air pollution sciences from intermittent inspection to ever more elaborate instrument-based surveys of the atmosphere. Since the first concerns with the corporeal limitation of smoke observers were recognised, and the tireless work of the CIAP commenced, there has been a commensurate desire to relocate the monitoring of air pollution around approved technological devices. While offering a purportedly more objective location for the science of air pollution monitoring to operate, however, the monitoring device (whatever it may be) ultimately served to strip the air sample from its ecological reference points – simplifying the air for governmental calculation. Since the mid-1980s it has been possible to detect a shift in the geographical and technological location of air pollution monitoring science. The first shift has been spatial, as air pollution monitoring has moved beyond cities to incorporate a range of peri-urban and rural locations like the field in Sutton Bonnington described at the beginning of this chapter. The second change has involved the increasing use of ecological ersatz as replacements for the technological devices used for emission monitoring.

The UK's *Acid Deposition Monitoring Network* (ADMN) provides one of the clearest examples of the geographical relocation of government-supported atmospheric pollution monitoring in the 1980s. The ADMN was established in 1986 and is dedicated to monitoring the acid content of rainfall throughout the UK.[8] The ADMN emerged in direct relation to the national demands placed on the UK in the wake of the UNECE Convention on Long-Range Transboundary Pollution. At one level there appears to be nothing significantly new about this network: the British government has long supported the monitoring of acid-producing air pollutants and associated rainwater analysis since the CIAP and the standard deposit gauge in the 1920s. What marks the ADMN from the National Air Pollution Survey (or its own predecessor, the *Sulphur Dioxide Monitoring Network*, SDMN), is that it is not so much concerned with how much acid-forming pollutants enter the atmosphere, but how these pollutants move through the ecological vectors of the air and affect interconnected environmental systems. Accordingly, the erstwhile Department for Environment, Transport and Regions (DETR) stated that, 'The deposition data from the network [ADMN] provides the foundation to the DETR-funded research programme that attempts to determine how acid rain is affecting sensitive ecosystems in the United Kingdom (DETR, 2001: iii). Originally the ADMN comprised 68 monitoring stations, but in 1989 this was reduced to 32 locations (ibid.: 3). In 1999, however, an additional seven locations were added to the network in order to cover ecological areas that were deemed to be particularly vulnerable to acid depositions (ibid.: 4). The location sites for ADMN operations now include remote locations such as Bannisdale in the Lake District, the Cow Green Reservoir in the Northern Pennines, the Strathvaich Dam in the Scottish Highlands, and Llyn Llagi in Snowdonia.

The ADMN utilises a number of bulk rainwater sampling devices in order to collect weekly, or fortnightly, atmospheric readings at its different sites. Since its inception, the ADMN has gradually incorporated a range of other monitoring devices for recording atmospheric concentrations of other pollutants responsible for acid rain (including nitrous oxides, particulate sulphates and sulphur dioxides, inorganic ammonium, and other acidic gas and aerosol species) (ibid.: 4). Collectively, these various devices, located in diffuse rural locations, provide something in the region of 10,000 sample measurements of acidic atmospheric deposits in Britain every year.[9]

Looking geographically at the operation of this new monitoring science it does appear to reflect what Luke (1999) would describe as an ecological coding of atmospheric knowledge production in Britain. While the scientific procedures of the network do not differ significantly from previous, biopolitically focused monitoring apparatuses, it is clear that the relocation of this atmospheric science reveals a desire to understand what the impacts of air pollution are on the non-human spaces of upland ecology and water systems. While it would be analytically crude, in the extreme, to suggest this geographical relocation of science can be used as a definitive indicator of a new ecological governmentality, it is clear that the types of knowledge that the ADMN produces makes it possible to think of atmospheric government in more ecologically conditioned terms. These ecological terms of reference are not based upon a new level of awareness of the actual ecological damage that species of acidic pollution are causing in different environmental settings (this is not, after all, analysed by the ADMN). Instead the ecological rationalities that are supported by networks like the ADMN can be discerned in the institutions and groups that utilise these newly coded knowledge sets. In the case of the ADMN, the knowledge sets it has produced were initially used in the expert deliberations conducted by the government's *Review Group on Acid Rain* (RGAR). More recently the ADMN has fed its results into the integrated *National Expert Group on Transboundary Air Pollution* (NEGTP) (ibid.: v). The ADMN also constitutes an important territorial cog within the *European Monitoring and Evaluation Programme*. The formation of governmental groups dedicated to the study and abatement of the territorial spread of transboundary pollution such as acid rain reflects the growing realisation that the atmosphere is both a context for non-human life and a complex ecological vector for the spread of air pollution into other terrestrial and hydrological systems. (An interesting quality of acid rain is that while the pollutants that lead to its formation derive from centres of industrial activity, the way in which acid rain is formed and transported in the atmosphere means that it often poses its greatest threats in non-human environments.) The partial relocation of the sites of atmospheric monitoring from metropolitan sites of air pollution production to non-urban sites of deposition thus enables scientists to trace the flow of effluvia though complex ecological systems. It is at the intersections of ecological knowledge and governmental power like these that it is possible to discern the rise of a new ecological governmentality within British air government. In this particular context, however, it is important to recognise that ecopower is not just evident in governmental care for new ecological spaces, but also in the clear recognition that governing the environment requires a holistic ecological mindset and a knowledge-gathering apparatus to match.

The geographical relocation processes encapsulated within the ADMN are, however, only one of two relocation processes that appear to be connected to the eco-coding of the apparatus of air knowledge production in contemporary Britain. Increasingly, various forms of government-sponsored air monitoring research are being relocated from the manufactured diffusion tubes and bulk collectors of networks such as the ADMN, and onto direct ecological indictors for the measurement of air pollution. While the UK currently monitors air pollution through a series of biotic and ecological ersatz, the National Moss Survey represents one of the most extensive networks of biotic indictors, and provides an interesting insight into the governmental rationalities and scientific techniques that surround these surveillance practices. Mosses have been used for some time as reliable biotic devices in and through which to measure the presence of metals in the atmosphere. The first extensive use of mosses to measure trace metal pollution was conducted in Nordic states during 1980 (Ashmore *et al.*,

2002). During the 1990s a European network of moss monitoring emerged to which Britain contributed data in both 1990 and 1995. Rhüling and Tyler were the first to develop the scientific technique of using moss species as collection devices for deposited trace metals in the 1960s (see Rhüling & Tyler, 1968). They were concerned with developing an ecologically based approach to the measurement of deposited lead. There are very specific physiological reasons why mosses are well suited to pollution monitoring work. First, mosses tend to extract most of the nutrients they require for survival from deposited and absorbed atmospheric sources (Ashmore *et al.*, 2002). In addition, the particular physiological form of moss cuticles and roots enables them to maximise the intake of atmospheric compounds while minimising the amount of absorbed terrestrial sources (ibid.: 6). These characteristics, combined with their wide geographical spread, make mosses ideal biotic tools for atmospheric sampling.

In this section I focus upon a national survey of mosses that was conducted in Britain during the summer of 2000 by a group of university-based scientists working on behalf of the British government. The British Moss Survey of 2000 involved 210 sampling locations that were spread throughout Britain and focused on four main moss species: *Pleurozium schreberi*, *Hylocomium splendens*, *Hypnum cupressiforme* and *Rhytidiadelphus squarrosus*.[10] The moss survey was governed by very strict *protocols of analysis* that involved the collection of one litre of moss, and its treatment, under heat, with ionised water and nitric acid. Once treated the moss samples were analysed for traces of calcium, potassium, magnesium and sodium using plasma atomic emissions spectrometry, and arsenic, cadmium, chromium, copper, nickel, lead, selenium, vanadium and zinc by plasma mass spectrometry.[11]

What is striking when looking closely at the operation of the 2000 Moss Survey is that the use of moss as an analytical indicator of air pollution does not in itself reflect an ecologically informed governmentality. In some ways the use of moss in the monitoring of metal pollution can be interpreted as a kind of biotic technology that directly replaces anthropogenic devices without necessarily implying a shift in the governmental intent of science. The point is that the presence of moss within the atmospheric knowledge-gathering apparatus of the British State does not directly denote a new concern with the ecological impacts of trace metal pollution. Indeed the presence of moss as an ecological entity in governmental surveillance is far from a clear indicator of ecological care in government: the selection of moss species is, in part, based upon their relative resistance to the short-term impacts of air pollution, while the disruption of moss habitats induced by the collection of litres of moss from 185 sites is not typical of systems of environmental conservation. As with the work of the ADMN, the connection between the National Moss Survey and an ecological governmentality stems not so much from the scientific techniques it involves, but from the geographies of its operation. The locations used as collection sites in the survey vary from urban parks and gardens, to cow pastures, and more remote lochs and moor lands. The varied locations appear to indicate a clear mixing of both biopolitical and ecological governmentalities within the programme. In this context, it is clear that the presence of moss analysis in urban locations reflects a desire to understand the relative concentration of trace metals in areas of high population – a position from which it is possible to extrapolate various health implications. In a related sense, the use of moss indicators in agricultural areas provides a framework against which to understand how trace metals may affect crop yields or enter the human food chain. It is thus, only when moss is located within remote ecological systems – that are divorced from immediate concerns with human health and food economies – that the survey appears to either reflect or facilitate an ecological reason for government. The subtle mixing of biopolitics and ecological governmentalities within the survey serves as an important reminder of the need to remain aware of the constant overlap – and subsequent analytical slippage – that exists between these two regimes of atmospheric government.

The age of ecological security: assessing the environmental thresholds of the atmosphere

A second way in which it is possible to discern an ecological mentality within contemporary British atmospheric government is in relation to emerging systems of ecological security. As previously discussed in his analysis of the apparatus of security, Foucault claimed that modern governmental practices are characterised not by a desire to impose an idealistic blueprint upon a nation, but by the continual calculation of *bandwidths of the acceptable* (see Chapter Two) (2007 [2004]: 6). While Foucault largely charts the establishment of the thresholds of government in relation to key biopolitical considerations (including food supply, disease and sanitation), it would seem reasonable to assume that if an ecological mentality of government existed it would involve the establishment of bandwidths of ecological, as well as social, acceptability.

Perhaps the clearest example of the governmental science of ecological security can be discerned in the formation of a British system of critical load analysis. A critical load is defined as, '[a] quantitative estimate of an exposure to one or more pollutants below which significant harmful effects on specified sensitive elements of the environment do not occur to present knowledge' (Centre for Ecology and Hydrology, 2008). The practices of critical load analysis and assessment originated within scientific studies of the processes of sulphur loading associated with acid rain (O'Riordan, 1999: 102). According to O'Riordan, the notion of a critical load has important governmental implications because it, 'presumes that the science [of pollution accumulation and transmission] is sufficiently well known to be used as a guide for policy' (1999: 102). As an emerging assessment procedure, critical load analysis was a scientific product of the frameworks of international cooperation and policy development that were established in Europe in the wake of the 1975 Helsinki *Conference for Security and Cooperation in Europe*, and in particular the monitoring programmes associated with the LRTAP (see above). In essence critical load analysis provided the scientific basis for assessing the impacts of such transnational policies.

The first connection between the LRTAP and critical load analysis was established in 1980 with the formation in Europe of the *Working Group on Effects of the Convention on Long-Range Transboundary Air Pollution*. This was an important institution because it provided the context for the construction of new forms of social and ecological data sets concerning air pollution. As its name suggests, the working group sought to support the LRTAP not only by monitoring critical levels of atmospheric pollution, but also by collating knowledge on the effect of air pollution on different geographical locations and environments. The coordination of scientific studies into critical loads of air pollution in Europe was formalised in 1988 when the European *Working Group on Effects* formed the *Task Force of Mapping*, which was eventually renamed, rather cumbersomely, the *International Cooperative Programme on Modelling and Mapping of Critical Loads and Levels of Air Pollution Effects, Risks and Trends*. More recently, the idea of critical loads has become the basis for the formation of new atmospheric protocols operating under the expanding remit of the LRTAP. In 1999, for example, the UNECE supported the formation of the Gothenburg Protocol (to abate acidification, eutrophication and ground level ozone), which sets *emissions ceilings* for air pollution.[12] While these emissions ceilings in part pertain to health considerations, they have also been designed to recognise the effects of air pollution on different environmental systems. In essence the Gothenburg Protocol is an international political framework for atmospheric pollution abatement that seeks, for the first time, to regulate pollution not only on the basis of its concentrations in the atmosphere, but also in terms of its potential socio-ecological impacts.

Although the science of critical load analysis is a product of a European-wide policy collaboration it has gradually assumed a prominent position within British systems of atmospheric government. The *UK National Focal Centre for Critical Loads Modelling and Mapping* (UKNFC) has scientific responsibility for the formation, mapping and monitoring of critical loads for atmospheric pollution. The UKNFC is based at the Centre for Ecology and Hydrology's offices at Monks Wood in Cambridgeshire and is funded

Table 8.1 Ecosystems for which critical load data are currently calculated

Ecosystem types
Coniferous forest
Deciduous forest
Mixed forest
Unspecified forest
Mediterranean forest
Acid grassland
Agricultural grassland
Alpine grassland
Calcareous grassland
Natural grassland
Grassland/reed/marsh
Heath
Heathland
Moors and heathland
Semi-natural ecosytem
Semi-natural vegetation
Tundra
Lake
Lake/stream
Freshwaters
Alpine lakes
Bog
Oligotrophic bog
Other

Source: Hall *et al.* (2001: 7)

by the Natural Environmental Research Council, Defra and the devolved administrations. While providing the scientific framework within which different European States operate, the *International Cooperative Programme on Modelling and Mapping of Critical Loads and Levels of Air Pollution Effects, Risks and Trends* allows individual countries to have responsibility to assess and designate critical loads for different ecosystems and air pollution types (Hall *et al.*, 2001). Table 8.1 reveals the different types of ecosystem for which scientists working throughout Europe currently calculate critical loads.

The analysis of critical loads in Britain involves two processes: (i) the mapping of main habitat types; and (ii) the calculation of critical loads for these different ecosystems (Centre for Ecology and Hydrology, 2004: 5). The Centre for Ecology and Hydrology deploys two methodologies in order to determine critical loads. The first, the mass balance approach, involves the detailed chemical analysis of the flows of key nutrients in and out of a given ecosystem. The second, the empirical method, relies on existing studies of habitat types and associated assessments of the levels of pollution that lead to the disintegration of given ecosystems (ibid.: 6). Once habitat zones have been mapped and critical loads calculated it is possible to utilise measures of air pollution to form what are intriguingly entitled *maps of exceedance*. Maps of exceedance reveal where levels of air pollution are leading to the crossing of key ecological thresholds and to the potential damage of key habitat spaces. The maps of exceedance produced by scientists working at Monks Wood provide an interesting cartography of ecological security – a geographical reference grid upon which air pollution abatement policies can be assessed directly in relation to their effectiveness in preserving certain forms of ecological vitality.

The new apparatus of environmental knowledge associated with the UKNFC appears to provide a clear indication of the ecological calibration of air pollution science in Britain. Unlike the Moss Survey, scientists at Monks Wood do not utilise fragments of ecology to test air quality, instead they use air pollution data as a basis to assess ecological resilience. The key to understanding this changing governmentality is the distinction that the UKNFC makes between *critical levels* of atmospheric pollution and the *critical load* of habitats. While a critical *level* of atmospheric pollution suggests a particular, and isolated, quality of the aerosphere, a critical *load* provides a way of ecologising the atmosphere and recognising its differential relationship with various ecological systems and processes. Crucially, this shift in scientific method also belies a change in the types of knowledge-gathering practices that are associated with air pollution science and government. Suddenly, as much, if not more, attention is given to the reconnaissance of various forms of information about habitat type and form as it is to actual levels of pollution in the atmosphere. Despite the role of critical load science in the ecological coding of air pollution, caution must be taken in equating it with a form of government that exhibits an all-embracing ethos of environmental care or *ecophilia*. If we look more carefully at the impacts of the science of critical load analysis on systems of atmospheric government in Britain it appears to have facilitated a highly strategic and disciplinary form of governmental power. One of the key assumptions of critical load science is that it is possible to establish a threshold of ecological tolerance, below which atmospheric pollution is deemed to be acceptable and sustainable. The idea that it is possible to draw such definitive lines of tolerance within nature is undoubtedly why the critical load models have proved so popular within both European and British air government (see O'Riordan,

1999). Critical load science therefore not only constructs data sets of habitat ecology, it also presents an ontological view of a steady-state nature that is governable alongside certain socioeconomic needs for pollution. Yet it is precisely such steady-state views of nature that are coming under increasing scrutiny from the new strands of ecological sciences of the late twentieth and early twenty-first centuries (see Botkin, 1992). Can we really know what the effects of an *acceptable load* of air pollution are on an ecosystem over a long period of time? Furthermore, even if acceptable levels of air pollution may not completely undermine the functional integrity of an ecosystem in the short or long term, is it not inevitable that it will erode the complex intricacies of ecological and biological diversity, and replace them with an ecological realm that is only secure on the narrowest of scientific terms?

Umbrella programmes, flexible thresholds and integrated atmospheric research

Although recent developments in the sciences of air pollution monitoring and assessment in Britain are clearly indebted to the emergence of the holistic sciences of ecology, it is far less certain that these developments reflect or foster an ecological governmentality. Perhaps we should not find this situation surprising. As previously outlined, Donald Worster's work reveals that the history of ecology is not a neat narrative of a single disciplinary science, but a diverse 'penumbra' of interconnected science practices and beliefs (1994, x–xi). It is in this historical context that it is possible to appreciate that the labelling of atmospheric science as 'ecological' does not necessarily mean that it is grounded on an ethos of practice that is significantly removed from studies of the 'environment-as-human-resource' that have pervaded governmental thinking for many centuries. Consequently, in order to discover evidence of an ecological governmentality towards atmospheric relations in Britain it is necessary to return to the team of scientists working in those curious fields in Sutton Bonnington with whom this chapter began. What differentiates the work that is being carried out at Sutton Bonnington (and other allied sites) is that it is a science that is not only characterised by ecological objects of analysis, and locations, but also by the active deployments of ecological methodologies. Sutton Bonnington is a part of a network of scientific sites that constitute the British government's *umbrella programmes*. The idea of an umbrella programme is to use a variety of different terrestrial surfaces and ecosystems as a basis for measuring changes in the concentration and spread of different air pollutants. In this section I want to consider the scientific practices and associated reasons for government associated with the British government's Ozone Umbrella Programme (OUP), which partly operates out of Sutton Bonnington.

Ozone pollution in the troposphere is caused when oxides of nitrogen and volatile organic compounds (VOCs) (produced in the burning of oil and gas) react with oxygen in warm atmospheric conditions. Ozone pollution in the lower atmosphere is, of course, synonymous with the smogs of large urban areas and is routinely monitored by the British government's Automated Urban and Rural Network of air stations. Increasingly, however, scientists have become aware of the long distances over which ozone can travel and the damage that it can cause in non-urban environments. It is in this context that the OUP was set up to monitor and assess the impacts of ozone pollution in different environmental spaces throughout Britain. The OUP is conducted and maintained by ecologists working in the aforementioned Centre for Ecology and Hydrology. Interconnected research programmes operating out of sites like Sutton Bonnington explore the impacts of ozone pollution on grasslands, arable crops (including sugar beet, wheat and oats), woodlands and wetland areas (Centre for Ecology and Hydrology, 2001). At one level it is clear that the studies carried out by CEH scientists into the impact of ozone pollution on the productivity of sugar beet, wheat and oats, appear to reflect an, admittedly novel, biopolitical intent within government.[13] Related studies of the effects of ozone on beech trees and various wetland habitats do, however, seem to represent a clearer expression of *ecophilia* within the OUP. As was discussed previously though, I want to argue that the ecological targets of atmospheric monitoring should not been interpreted as markers of an ecological governmentality, rather that the nature of the scientific practices that surround such objects should be our primary concern.

Looking more closely at the scientific methods deployed within the OUP it becomes easier to discern an eco-coding within the governmental knowledge it is endeavouring to produce. The key measurement that essentially connects the different monitoring operations of the OUP is something called AOT40. AOT40 is a critical level of pollution employed by the British government to determine the thresholds of acceptable atmospheric ozone concentration above which plants are likely to suffer significant damage. AOT40 designates ozone pollution as, 'accumulated exposure over a threshold of 40 ppb' (Centre for Ecology and Hydrology, 2001: 29). As with the critical levels discussed in the previous section, AOT40 reflects an obvious example of emerging systems of ecological security and threshold assessment in British atmospheric government. What is, however, interesting about the work of the OUP is not so much its recording of the effects of different levels of ozone on various environmental spaces (as with critical loads science), but the way in which it has used ecological methods of scientific assessment to critically question the designation of government thresholds for air pollution. Through the different research sites operating as part of the ozone umbrella, scientists working within the CEH combined the use of ozone measurement devices with the monitoring of other atmospheric and environmental conditions, and the careful study of plant form and physiology (hence the curious mix of crops and instruments at Sutton Bonnington). In essence the OUP deployed what could be reasonably equated with the holistic scientific methodologies associated with ecological assessment. The value of moving beyond the type of reductionist monitoring methodologies that simply measure concentrations of pollutants in the atmosphere and read off from this apparently related environmental consequences – or set rigid ecological loads for pollution – is that the OUP has started to question the governmental utility of fixed thresholds for air pollution. The work of the OUP has revealed that critical changes of biomass occur much more quickly in some species of the same tree, plant or crop than others (ibid.: 29). Further analysis has also revealed that the water content of the soil and ambient carbon dioxide concentrations in the atmosphere can also render the validity of ecological thresholds unreliable. Finally, the research of CEH staff working on the relationship between beech trees and ozone pollution revealed that beech forests were particularly sensitive to ozone in the early phases of the growth season. (ibid.: 25).

Ultimately the research conducted as part of the OUP has led to concerted scientific demands for the establishment of more ecologically sensitive thresholds for atmospheric pollution. The flexibility that the OUP call for relates not only to variations in the relative tolerances of different types of the same species or habitats, but also the seasonal timing of the threshold indicators. The OUP is essentially attempting to generate a different system of ecological coding than we have so far encountered. The key point is that this coding is not based simply on the setting of ecological standards for air pollution or habitat resilience, or the measurement of air pollution through ecological proxies, but seeks to utilise the insights of ecological knowledge gathering as the basis for determining the environmental security apparatus of the State. While the types of flexible ecological thresholds and security indicators suggested by the OUP may be difficult to enforce at a governmental level, the work of CEH scientists is suggestive of the ways in which new ecological methodologies could provide opportunities for the emergence of novel forms of ecological governmentality in Britain. These new ecological governmentalities would not so much be based upon a concern with the environmental impacts of air pollution, but on the use of ecological techniques as a basis for government action and intervention. What is clear is that a regime of ecologically coded security would necessitate a much more nuanced system of governmental regulation of air pollution: with the generic policing of fixed thresholds giving way to a flexible system of intensive pollution control within bio-ecologically sensitive times and places.

Such systems may, for example, involve the enforcement of stricter standards of air pollution control at key times of the growing season, or recognise the need for the enforcement of different critical levels of pollution control within cities during different meteorological conditions.

Conclusion – Reflections on an Environmental Revolution

The idea that Britain has gone through something of an environmental revolution in government has been generally predicated on two sets of processes: (i) the rise of a new set of environmental institutions within British government (particularly the Department of the Environment, the Central Unit on Environmental Pollution and the Royal Commission on Environmental Pollution); and (ii) the connections forged between Britain and expanded systems of transnational environmental policy and surveillance associated with the European Union and United Nations. Focusing on the specific question of atmospheric government, in this chapter I have argued that a profound shift in governmental thinking and action associated with the idea of an environmental revolution cannot be reliably detected through an assessment of the institutional restructuring of government alone, but must be based, at least in part, on the presence, or absence, of new regimes of environmental/ecological knowledge production. Having explored some of the new systems of State-sponsored atmospheric science that have emerged in Britain since the late 1960s, my assessment of the significance of the environmental revolution is similar to Zhou Enlai's characterisation of the impact of the French revolution on political history: namely 'It's too soon to tell' (Schama, 2004: xv). My use of Zhou Enlai's famously laconic statement is not meant to justify a convenient intellectual side-step; it is just that considering the impacts of new modes of environmental thought and ecological practice on atmospheric government in Britain has left me with the firm belief that any revolution that may have taken place is still very much in progress.

The aim of this chapter was to address two interconnected questions. First, how would we recognise an ecological governmentality? Second, what, if any, forms of ecopower are characteristic of British atmospheric government over the past 40 years? In relation to the first question, this chapter has asserted that the notion of ecological governmentality should not be used as an ontological heuristic against which test and compare practices of government, but should instead, and in keeping with Foucault's own historical method, suggest a method of analysis for exploring the emergence of new rationalities for governing. To this end, the chapter has sought to reframe the question: it is not a question of how we would recognise ecological governmentality, but how we could begin to search for it. In keeping with the methods deployed throughout this book, this chapter has thus searched for traces of an ecological governmentality through an excavation of the apparatus of knowledge production associated with atmospheric government in Britain. This excavation processes has revealed that it was not until the 1980s, and the emergence of key international initiatives of air pollution assessment, that the purported environmental revolution in British government appears to have made a significant impact on the apparatus of atmospheric knowledge production. After this point, however, it is possible to discern the techniques of ecological science and scientists taking an increasingly important role in shaping how and why atmospheric knowledge is produced – or to paraphrase Rose and Miller (1992), the assimilation of ecological thought into the *intellectual machinery* of government.

This chapter has revealed that there are two ways in which regimes of ecological governmentality can be understood. The first sees the governmentalisation of ecological thought and science – or ecological systems and processes being subjected to the simplifying logics of minimalist neo-liberal government (becoming part of what Rutherford describes as the expanding 'governmentalisation of life') (Rutherford, 1999: 60). The second way of thinking about ecological governmentalities is in relation to the knowledge and practices of ecological science recalibrating the actions of government and ultimately generating more cautious regimes of environmental security and more modest rationalities of ecopower (what Rutherford terms the *institutionalisation of ecological rationality*) (ibid.: 59).[14] An analysis of the contemporary apparatus of atmospheric knowledge production in British government illustrates that while it is clear that the environmental revolution has produced the former mentality, it is far less certain that the later system has established itself within air pollution policy making. This situation makes it important to reconsider the relationships between ecopower and biopower discussed earlier. While suggestive of a new reason for government, it is clear that major systems of what appear to reflect ecopower, operating in and through the contemporary British atmosphere, are either extensions of a biopolitical knowledge-gathering apparatus, or tend to transpose the simplifying biopolitical logics of governing a population on to the techniques employed when governing ecology. Where more distinctive, and holistic, systems of ecopower (perhaps best understood as power with, rather than over ecology) are evident – as in the case of the OUP – it is clear that the apparatus of atmospheric knowledge production and security are far less extensive, consistent or well supported than their biopolitical counterparts. Consequently it appears that the contemporary atmospheric knowledge-gathering apparatus in Britain continues to be dominated by both biopolitical interest and rationality (Rutherford describes this persistent legacy as the 'biopolitical character of modern governmental rationality') (ibid.: 56). Thus where ecological governmentalities do exist they appear to be conditioned by regimes of biopolitical knowledge production, which means it is difficult to make an escape from narrowly defined anthropogenic concerns, and processes of governmental simplification, which a regime of ecological government appears to promise. To put it another way, it appears that in Britain there has been a sharp increase in the role of ecology in providing knowledge concerning air pollution and available environmental assessment techniques, but that this new epistemology of governmental knowledge has not been mirrored by the rise of a distinctive ecological rationality of government.

Chapter Nine

Conclusion: Learning Like a State in an Age of Atmospheric Change

To think historically is to see the world as always contingent, as an impure and imperfect product of human actions and environmental processes over time [...] one of the gifts of a historical education is knowing that some wounds heal in time or can be endured, and that we do not have to go it alone

Klingle, 2007: xii–xiii

Concluding Reasons and Reasons of Conclusion

This chapter offers a slightly unconventional form of conclusion. While it does provide the critical review of this volume that is customary within a conclusion, there are two further objectives. First, it utilises the history of air pollution science and government presented in the preceding chapters as a context within which to interpret the contemporary practices and controversies that surround climate change abatement strategies in Britain. As unquestionably the most prominent, and certainly keenly contested, areas of contemporary atmospheric debate, this chapter critically analyses how the history of British air science and government can help us to interpret, and challenge, the unfurling rationalities of climate mitigation in Britain. Second, it provides a degree of normative perspective on the analysis presented within this book. This normative dimension addresses the following issue: on virtually all counts and measures (and with the notable exception of greenhouse gases) the fusing of atmospheric science and government in Britain has resulted in more being known about the chemical composition of the atmosphere, declining absolute levels of air pollution and reduced levels of air pollution related health complaints within the population. In this context it is legitimate to ask, what is the value or purpose of a critical scientific and governmental history of air pollution? To put it more plainly, why be critical or suspicious of historical processes that appear to be ostensibly successful?

Drawing on recent work on civic science and collective learning, this chapter attempts to challenge the construction of a celebratory history of air pollution abatement in order to consider alternative ways of knowing and governing the atmosphere. This search for alternative regimes of knowing and governing is a crucial, if at times implicit, goal of Foucauldian governmentalities. Although Foucault's analyses of liberal forms of government have routinely been characterised as highly pessimistic accounts of power within Western society,[1] it is clear that when placed alongside Foucault's wider political activities, and his immersion in left culture, that his governmentality lecture series can be interpreted as a context within which to understand the potential for developing new (perhaps socialist, perhaps eco-socialist) governmentalities (see Chapter Two).[2] In part this analysis of alternative ways of knowing and governing the atmosphere emerges out of a consideration of the social and community utility of contemporary forms of atmospheric knowledge in Britain. At another level, however, this chapter attempts to marshal insights into the spatial and historical contingencies of air science and government as a basis for demarcating the practical and epistemological opportunities that exist for the mobilisation of new forms of air government with science.

Analytical Themes: Atmospheric Government in Critical Prospect and Retrospect

Spatial histories and 'cartographies of the present'

At the outset of this volume I ascribed my motivation for writing this book to a desire to understand the production of contemporary forms of atmospheric knowledge in relation to the processes and practices of air *archiving* (see Preface). This recourse to the practices of archiving was both practical and epistemological. At a practical level I was fascinated by the scientific and governmental rationalities that informed the seemingly incongruous production of Britain's *Air Quality Archive*. At a more epistemological level I was interested in the suggestion that is contained within the very notion of *air archiving* that time can become a framework of intelligibility through which we can come to know, predict and perhaps even control the atmosphere. But despite providing crucial intellectual stimulus for this project, my initial concern with the practices of *air archiving* has been replaced by a desire to understand the *spatial history* of atmospheric government and science.

In keeping with Foucauldian genealogies, my desire to construct a history of air pollution government was not so much concerned with the temporal ordering of air knowledge, but with the role of historical perspective in revealing the arbitrary nature of contemporary ways of knowing and being. It is in this context that Nikolas Rose asserts that,

> [g]enealogies seek to destabilise a present that has forgotten its contingency, a moment that, thinking itself timeless, has forgotten the time-bound questions that gave rise its beliefs and practices (Rose, 2007: 4–5).

My recourse to the historical study of atmospheric knowledge and government is thus more akin to an archivist trying to unlock the codes of an existing historical record than to the allocation of temporal order to an undifferentiated knowledge set. My more specific concern with spatial history – while in keeping with key aspects of Foucault's own approaches – was not born out of conceptual conjecture, but from the impetus generated by studying the practices and micro-politics of air science and government.

Much has been written on the overt and latent spatial dimensions of Foucault's histories (see Crampton & Elden, 2007; Elden, 2001; Philo, 1992).[3] While detailed studies have been conducted of Foucault's use of spatial metaphors of boundary and threshold within his early archaeological conceptualisations of knowledge, and his suggestive reflections on the value of developing a history of spatial forms, more recent analyses have emphasised the presence of *substantive geographies* within Foucault's main body of work (Elden, 2001: 93–119). According to Elden, these *substantive geographies* suggest more than merely a sensitivity to changing notions of space through time, but reflect an appreciation of the role of geographical extent, proximity, location and place within the formation, consolidation and contestation of knowledge and power (ibid.: 111–19). To put it a different way, *spatial histories* differ from *histories of space* to the extent that they '[u]se space itself as a critical tool of [historical] analysis' (ibid.: 119). A spatial history does not consequently just confront teleological history with contingency, but explicitly reveals the geographical contours and spatial practices that under-gird the conditional nature of all knowledge. Spatial histories do not simply add another dimension of contingency to those offered by historical analyses; they expose the role of geography in unsettling the certitudes of knowledge production and maintenance. To study the micro-practices associated with the governmental production of atmospheric knowledge has thus required an open immersion in, and concern with, the problems that the simultaneous gathering and coordination of data over large spatial areas and territories involves, and the political compromises it necessitates. To explore the practices of governmental knowledge production is to study a history that simply cannot afford to ignore space.

Space has been present within the histories of air pollution government with science presented in this volume in three main ways: (i) as method; (ii) as epistemology; and (iii) as rationality. The methodological manifestation of space within this book has perhaps been the most tacit geographical modality of all. Space has been manifest as method, however, in two main ways. First, this volume has been predicated on the study of archival records that have been deliberately drawn from wide geographical points of origin, extending from London to Glasgow, Manchester to Aberystwyth, and Birmingham to Exeter. While it would have been possible to construct a narrative of British atmospheric government with science that was based upon the central records of Parliamentary Committees, the Committee for the Investigation of Atmospheric Pollution, the Department for Scientific and Industrial Research, or the Warren Spring Laboratory, such records would not have revealed fully the contingent forces that are induced as scientific practice and governmental knowledge are made to travel through space (see Shapin, 1995). Second, and related to the first method, analysis in this volume has paid particularly close attention to the immutable mobiles (ranging from deposit gauges, instruction manuals, charts, new kitchen ranges, and even digital signals) that have enabled atmospheric government and science to move in space. Carefully following the travels of these different objects has facilitated insights into the role of geographical location, place and mobility in the constitution of environmental knowledge. The struggles for standard practice that surround these various objects have made it possible to discern the importance of both precise location and socio-cultural place within the formation of atmospheric knowledge. But the stories that surround these objects not only help expose the fact that government and science operate in contingent places (as opposed to an abstract, frictionless space of absolute science and government), but also emphasise that the ways in which atmospheric knowledge moves are also meaningful. It is in this context that I believe much can be gained by extending contemporary Foucauldian work on the geographical constitution of knowledge from studies of the role of culturally meaningful space (or *place*), to analyses of meaningful movement (or *mobility*) (see Cresswell, 2006). The tales of knowledge movement explored in this volume (including the accounts of smoke inspectors' journeys through city streets, travelling exhibitions, and even the connectivity on the World Wide Web) are crucial to the effective construction of the spatial histories of government and science.

The second modality of space that has run through the various chapters has been epistemological. In using spatial epistemology I am referring to the different ways in which the historical production of atmospheric knowledge is conditioned by space. While the spatial conditioning of knowledge production can be discerned in relation to the problems of maintaining extended networks of air monitoring described above, it has also been event in this volume in other ways. In Chapter Three, for example, we saw how the physical morphology of the urban landscape provided both opportunities for, and barriers to, the effective visualisation of air pollution by smoke observers. The links between space and epistemology took a very different form in Chapter Five. Here we saw how the birth of volumetric measures of air, and the construction of standard atmospheric regions, provided a rational basis for developing both a form of spatial science of the atmosphere and geometric frameworks within which to organise air knowledge. In this context, space became less a frictional force on the fashioning of atmospheric knowledge and more an explanatory context for air interpretation.

The final modality of space in this volume has been present in relation to questions of rationality. The association between space and rationalities (or the balanced reasons) of and for government has been evident in a form of dialectical relation. On the one hand the spatial location and directions given to atmospheric science have provided clues to changes in the nature of atmospheric governmentalities. Early in the volume, the redirecting of the smoke observer's gaze from the litigious concerns of prosecutions to a more general interest in the conditions of the urban air revealed an important extension of disciplinary forms of air government into the realms of atmospheric security. In a similar sense, at the end of the Chapter Eight analysis of the evolving relationships between forms of bio- and ecopolitics revealed the importance of the location of science when assessing the extent to which the rationalities of air government have actually been transformed in this new environmental age. On the other side of this dialectical relationship, however, we have also seen how, at various stages in the development of British atmospheric government, spatial perspectives have not only mirrored governmental rationalities, but also actively shaped them. The operation of the National Air Pollution Survey during the 1960s, for example, revealed how a belief in the spatial patterning of atmospheric relations informed new patterns of geographically calibrated policies for pollution abatement that focused on housing quality and home heating cultures (particularly in the north of Britain). In a similar way, we also saw how the new technological ability to produce ever more elaborate cyberspatial emulations of the atmosphere, through the National Atmospheric Emission Inventory, has facilitated a more flexible rationality for governmental intervention in atmospheric relations.

If the three modalities of method, epistemology and rationality provide the basis upon which this volume can be conceived of as a spatial history of atmospheric government with science in Britain, I want to conclude by considering the utility of such a geographical history. At the start of this section I reflected upon how historical genealogies, as envisaged by Foucault and his subsequent acolytes, tend to destabilise the apparent inevitability and *solidity of the present* to reveal other ways of knowing the world and being within it (Rose, 2007: 5). Rose has recently challenge the role of genealogy by suggesting that our analyses should be focused more on developing *cartographers of the future* as opposed to histories of the present. Rose asserts that, 'Such a cartography would not so much seek to destabilise the present by pointing to its contingency, but to destabilise the future by recognising its openness' (ibid.: 5). While Rose tends to deploy the notion of cartography in a somewhat metaphorical fashion, arguing for the need to develop, '[a] map showing the range of paths not taken that may lead to different potential futures' (ibid.: 5), I want to argue that much can be gained by developing spatial histories that serve cartographies of future. Space, whether understood as the location sites of knowledge production, the distances that science and government must travel, or the stimuli of rationality, provides crucial clues to the practical opportunities that exist for redirecting the ways in which atmospheric knowledge is produced and the associated contexts in which we think about and regulate our relations with the air (I will reflect further on this project in the final section of this chapter, which discusses matters of collective atmospheric learning).

Entanglements of State and science: the thresholds and dialectics of government with science

The second key analytical theme that subtly, and at times more abruptly, weaves its way through this volume is the ever-contested relationship between the State and science. Although, in keeping with a Foucauldian perspective, this book has explicitly focused on the relationship between the practices of atmospheric government and the scientific modes of knowledge production through which such practices are sustained, recourse to 'the State' has never been far away. The latent presence of the State throughout the volume should not, however, be interpreted as a failure to remain focused on the micro-politics and practices of air government. As Foucault observes, adopting a concern with the micro-practices of political action and change does not necessitate abandoning the State as an object of analysis, but merely a refusal to use the State as an historical (and/or spatial) *a priori* mode of explanation. As a key institutional and territorial fulcrum of governmental power and practice States must inevitably appear in any discussion of atmospheric government. Moreover, as Timothy Mitchell so pertinently reminds us, the power (or *effects*) of the State can be discerned in the mundane assumptions that routinely divide the world into State-market, State-society, State-science dualities, and the patterning of action that derives from such belief systems, as much as it can in the material actions of any actually existing institutional leviathan (Mitchell, 2006).

The unavoidable squeamishness with which we approach discussions of the State and science derives from a paradox that was explained most famously by Immanuel Kant in his discussions of the constitution of the modern university (see Smith, 2004: 235–69). According to Kant there is a necessary dialectic existing between the realms of State and science. If science is to attain its destined objectivity (and associated authority) it must remain functionally independent from the State and the corrupting influences of associated political ideology and influence. At one and the same time, however, who or what can sustain science with the resources it requires, while preventing it from becoming the servant of special interests, if not the State (ibid.; see also Blissett, 1972: 11–25)? In the context of the apparent necessary interdependencies of the State and science, and the contradictory desires to clearly demarcate the political terrains of State action and scientific practice, it is important to consider what the history of British atmospheric government can tell us about the terrains of State science relations.

The defining encounter between the State and science described in this volume came in the guise of the 1843 Parliamentary *Select Committee on Smoke Prevention* (see Chapter Two). This Select Committee explicitly sought to bring together the apparatus of State with the 'great men of science' in order to address more effectively the problems of heavily polluted atmospheres in Britain. Yet, in many ways, the idea of this initial (formal) encounter of the State and science in the quest for cleaner air is misleading. The idea of a Parliament calling for the expert advice of scientists suggests the existence of two already demarcated zones of political and scientific existence, united only temporarily in the necessary pursuit of a common good. Not only is this idea of the initial enmeshing of State and science somewhat misleading, however, it also fails to reflect the nature of the relations that were eventually forged between the institutions of government and air sciences in Britain. The point is that science did not just offer a series of pragmatic technological solutions to the problems of air pollution production and monitoring (although this exchange certainly took place), but rather that the principles of scientific measurement and knowledge calibration provided a framework for the constitution of air government itself. This process was, admittedly, connected to the broader governmentalisation of the British State, but it is informative to briefly reflect upon the local practices of this shift through the example of atmospheric government with (air) science. What is particularly interesting about the scientisation of the State in Britain is that it brings into even more intimate focus the paradoxical dialectic of State–science relations, and the impacts of this dialectic on the actual practices of air government.

The practical contradictions of mixing State and science in Britain found corporeal form in the work of early smoke observers (see Chapter Three). In our discussions of the overworked atmospheric/nuisance inspectors who patrolled the streets of British cities during the nineteenth and early twentieth centuries we saw the combined demands of the State and science falling heavily on their shoulders. As agents of air government these inspectors were initially expected to adopt a sense of scientific objectivity in their assessment of specific air pollution events. And yet this ethos of science was undermined by provisions of resource that made systematic atmospheric surveys unthinkable; their commitment to the production of legally viable proof of air pollution as opposed to recording actual levels of air pollution; and the frail limitations of their own binocular bodies to deal with the ever increasing chemical complexities of urban air pollution. In essence, the bodily (in) capacities of smoke inspectors emphasises that, despite the desire for a system of government with science, the objectives of government (in this case the equitable enforcement of atmospheric law) are not always compatible with the systematic objectivities associated with scientific observation.

Beyond the tensions identified between atmospheric proof and truth, however, the historical records of smoke observers also reveal further discords between the desires of the State and science. These discords revolved around the fact that State officials, responsible for the day-to-day government of the atmosphere, were routinely caught between a scientific commitment to objective observation and a governmental commitment to paternal supervision and guidance. While modern environmental bureaucracies make it much easier for State authorities to institutionally separate out acts of scientific knowledge production from the forms of political intervention that utilise such knowledge, in the formative years of air pollution government in Britain such a separation of powers was not evident. Once again it was thus the smoke observers who had to reconcile the practical and ideological tensions associated with a purportedly more scientific style of air government. It was precisely in such a context that we saw smoke inspectors developing practices of *sub rosa* surveillance that would enable them to execute objective observation of air pollution one day, while on another exploit their close working relations with factory owners in order to encourage appropriate smoke abatement procedures. This practical tension between science and supervision (perhaps a defining characteristic of governmentality itself) found a more formal mode of articulation within the official selection and training of atmospheric inspectors within local areas. The joint recruitment of atmospheric scientists (specifically with meteorological training) and coal officers (who had more intimate knowledge of the workings of boilers and the needs of stokers) reveals an initially confused, but ultimately necessary, division of scientific labour within atmospheric government.

If the initial challenges of a more scientific brand of atmospheric government to prevailing ideologies of the State were experienced at a very local, even corporeal, level, it was not long until they started to affect more strategic forms of governmental decision making (see Chapter Six). In the mid-1920s two processes came together to re-problematise the relationship between the British State and air pollution science. First, it became evident that the Advisory Committee on Atmospheric Pollution (ACAP, previously known as the CIAP) required extra government resources if it was going to be able to expand its monitoring activities. Second, government restructuring at the time led to uncertainty over which department, if any, should have operational responsibility for the work of the ACAP. The simultaneous need of the ACAP for more State support and a home in government triggered a reassessment of the relationships between British air pollution government and science. This reassessment process was predominantly conducted through the hastily formed *Committee of Enquiry into the Future of the Advisory Committee on Atmospheric Pollution*. Two messages clearly emerged from this Committee concerning the perceived relationship between the State and science. The first was the belief that the State could only support the highest calibre of independent scientific research (the work of the CIAP/ACAP was seen as both amateurish and partisan). The second message suggested that the British State, with its duty to the responsible use of public money, could not support speculative research (such as that associated with the ACAP), but had to assist science that had clear, practical and commercial benefits. While the ideological division of the responsible, democratically accountable State, and the partisan, specialists of science is an unfair dichotomy (there is a long historical record of States, including Britain, supporting highly speculative research when it reflected the particular interests of government elites), this ideological distinction had institutional effects. In order to appease such concerns, and to allow the ACAP an institutional space within the State, the Committee was divided into the funded *Atmospheric Research Pollution Committee*, and the unfunded *Standing Conference of Cooperating Bodies*. While making the work of the ACAP a more acceptable destination for State support, this act of division ultimately created a highly fragmented and under-resourced system of air pollution government, which was incapable of dealing with the pressures that the 1952 London fog disaster ultimately placed upon it.

A much later process of government review outlined in this volume revealed a further aspect of the tensions between State and science in matters of air pollution government. In 1971 the *Interdepartmental Committee on Air Pollution Research* started a review of the yet to be published National Air Pollution Survey. Depicted as clumsy, outdated and unscientific, many top scientists were advising the government that it needed to invest in new, smart air monitoring equipment if it was to keep up with developments in North America and Europe. While, at one level at least, it is interesting to see how a national State project of atmospheric science could come under criticism for not being scientific enough, what is perhaps more important about this review process are the reactions that the suggested restructuring to Britain's air monitoring apparatus generated within government. During the early 1970s, leading scientists in the British government were calling for the concentration of State investment into a small number of highly sophisticated, state-of-the-art atmospheric monitoring stations. It was argued that the reliability and comparative advantages of such stations would provide far more useful scientific data on air pollution than the haphazard system of surveillance that had been used during the National Air Pollution Survey. The opposition to these downscaling proposals is, perhaps, indicative of enduring ideologies concerning the role of the State within society. Medical experts claimed that a smaller-scale air monitoring network would compromise the ability of the State to effectively calibrate pollution data with medical records. In effect, it was argued that to abandon large-scale atmospheric surveys would be to betray the social contract of the State towards the biopolitical welfare of the nation. Beyond medical concerns, further resistance was made to the downsizing suggestions on the grounds that such a spatially uneven network of surveillance would undermine the State's long-standing commitment to the territorial welfare of the population as a whole. In essence, the debates surrounding the Interdepartmental Committee on Air Pollution Research reveal that the ideological distinctions between the purported goals of the State and science are partly manifest in the spatial scope and territorial extent of knowledge production.

It is possible to distil the evident tensions between the practices of the State and science identified in this volume into two broad areas. First, there have been evident concerns over the continuing ability of the State to fulfil its functional duties (namely the enforcement of atmospheric law, and the provision of territorially comprehensive care) alongside a more scientific approach to air government. Second, the history of modern air government in Britain is clearly marked with an ongoing struggle to determine the appropriate modalities of scientific practice the State should support (this has been particularly evident in the debates over speculative and practical sciences). Such complex tensions appear to require the development of a more nuanced language when discussing State–science relations. What has actually been charted in this volume is not the general coming together of the State and science, as part of the universal governmentalisation of the State, but a series of tensions emerging over very specific expressions of both the State (its bureaucratic form, territoriality, supervisory logics and legal dynamics) and science (including the autonomous speculation of individual scientists, emergent forms of civic science, government-sponsored scientific experiments and government-orchestrated scientific projects). This situation appears to require a language that is able to analyse the complex and overlapping relationships between these different species of State and science. This new language should not, however, denote an acceptance that the notions of the State and science do not matter. What analysis has shown is that the history of atmospheric government in Britain is not something that can be explained by easy reference to the causal powers of either the State (in keeping with Foucault) or powerful scientists (following Latour). Notwithstanding this crucial conclusion, it is also critical to recognise (after Mitchell, 2006) the *effects* that the ideological constructions of singular and homogeneous entities labelled 'the State' and 'science' have had in ordering and directing atmospheric government in Britain. Extending Mitchell's framework of analysis, I would argue that the history of British air pollution government has been a history shaped by both *State effects* and *science effects*. While these effects cannot hide the complex interdependencies of State and science, they have, at different times, been used to shape the legitimate extent and moments of (in)action of both governmental and scientific communities in the government of atmospheric pollution. In a contemporary world, where the pressures for State and scientific action in atmospheric affairs has never been stronger, understanding how the artificial construction of the autonomous zones of State and science shape emergent forms of atmospheric action, and senses of responsibility for air government, has never been more important.

Conduct, self-discipline and the atmosphere subject

The third, and final, analytical theme that permeates various aspects of this volume pertains to issues of atmospheric conduct and the production of governmentally inscribed atmospheric subjects. In some ways it could be construed as strange to even talk about atmospheric subjectivity. From our first breath to our last we are all subject to the atmosphere and bound to it through a series of intimate relations. Yet much of our biological subjection to the air occurs at a pre-cognitive and mostly subconscious level (we do not, in ordinary circumstances at least, need to remember to breathe). It is precisely in this context that recognition and regulation of our broad socio-economic relations with the atmosphere appears to depend on a mixture of power and imagination that is tied into the history of government (see Agrawal, 2005: 199). To put things slightly differently, what this volume has shown is that while we may be acutely aware, at a corporeal level, of changes in the quality of the air that we routinely breathe, our ability to understand our role in producing and abating various forms of atmospheric pollution appears to depend upon the mobilisation of certain *technologies of imagination* (ibid.).

There are particular physical qualities of the atmosphere that, in marked contrast to other forms of environment (including land, forests and hydrological systems), make its government particularly dependent upon the technologies of the imagination. First, the ephemeral nature of the atmosphere tends to result in the actions that produce air pollution and contamination being removed from the spaces and places where they take worst effect. Second, the continual chemical mixing and bio-physical mulching that characterise the air, often make it difficult to differentiate the action of an individual polluter from the collective field of contamination (the smog is always someone else's fault). Governing the atmosphere is then, in part, a process of developing structures of the imagination in and through which people can better understand their own role in the collective production of the atmosphere. Consequently, throughout this volume I have charted not only how science and government have coalesced to produce knowledge of atmospheric pollution, but also how this knowledge has been channelled into the production of new regimes of air conduct and novel forms of atmospheric subjectivity. The associations between atmosphere, government and subjectivity traced throughout this book have taken four primary forms: (i) the use of the atmosphere as a medium for the regulation of social pathologies; (ii) the production of modes of atmospheric conduct deemed necessary to support the production of scientifically inscribed air knowledge; (iii) collective retraining programmes that have sought to transform individuals' relations with the air; and (iv) the production of the self-regulating atmospheric self. I want to consider each of these forms of atmospheric subjectivity in turn.

In his lucid analysis of environmental subjectivities, Arun Agrawal recognises the 'productive ambiguities' associated with the notion of a subject (2005: 165). Agrawal recognises that subjects can be seen as creative agents of social change and action (ibid.). At the same time, however, he also acknowledges that to be a subject is also synonymous with a more subservient vision of citizenship, whereby the individual becomes the object of power (ibid.). Rather then perceiving such ambiguity to be problematic, and in keeping with Agrawal's own analysis, I claim that positioning atmospheric subjectivities at the intersection of personal freedom and subjugation represents a crucial interpretive step when studying the connections between governmental power and the individual (see also Rose, 1999b: 40–7). In this broad context, one set of perspectives offered on atmospheric subjectivity in this volume is clearly indicative of the use of the atmosphere as a potentially oppressive social force. Many of the efforts of early atmospheric reformers in Victorian Britain were motivated not by pure paternal benevolence, but by a belief that the declining quality of the atmosphere was directly connected to the deterioration of the moral fabric and economic productivity of society. Much has already been written on the associations in the Victorian mind between various forms of social delinquency and the parlous state of the urban environment in the nineteenth century (see Driver, 1988). It is clear in this context that the urban atmosphere became a key medium for nascent governmental strategies for individual reform within the city. The emergence of smoke inspectors and allied sanitary authorities in nineteenth-century Britain was obviously a response to the evident health threats posed by polluted atmospheres, but they were also part of a broader apparatus of urban social reform. By improving the quality of the air urban authorities hoped to address a range of public policy concerns ranging from crime rates, juvenile delinquency and poor economic productivity.

According to Golinski, the amelioration of air pollution during the eighteenth and nineteenth centuries was part of a broader programme of atmospheric therapeutics through which various qualities of the air (including pressure, moisture and contaminates) were connected to the qualities of British civilisation itself (2006: 137–69). To these ends the atmosphere itself became a subject of government, which when properly ventilated could secure a socially docile, but economically energised, populace. While various fashions of environmental reform became associated with the curing of social pathologies in the late Victorian and Edwardian eras, it is worth reflecting on the particular role of the atmosphere as a tool, or subject, of government. If government, as Foucault suggests, involves the simultaneous ordering of the population and individual conduct, the atmosphere appears to afford qualities that enable it to serve the rationalities of governmental power in unique ways. As something that is shared as a constantly mobile resource, but intimately connected to each breathing subject, the atmosphere provides a potential site for collective forms of governmental action to operate concurrently on a population and a subject. Understood as a vector of power, which quite literally moves through multiple subjects, effectively governing the atmosphere appears to promise the goal of caring for 'all and each' (*omnes et singulatim*), which Foucault identifies as a key marker of governmentalities (Foucault, 2000c [1979]).[4]

The second aspect of atmospheric subjectivity uncovered in this volume pertains to the forms of conduct that became associated with those responsible for the collection and calibration of governmental knowledge about the nature and extent of air pollution. As described in the previous section, the emergence of modern forms of atmospheric government in Britain was in part predicated on the production of new forms of air inspectors. We saw how the dual role of air inspectors, as nascent air scientists and servants of local and national government, produced decidedly schizophrenic forms of subjectivity. It is important to recognise the connections between governmentality, subjectivities and knowledge production. It is clear that in order to produce the forms of knowledge upon which governmental forms of power depend, governments must initiate an intensive process of re-subjectification that centres on servants of the State.

While not wishing to repeat the forms of behavioural calibration that are associated with the scientisation of the knowledge-gathering capacities of States (see though Edney, 1999; Scott, 1998), it is clear that the emergence of modern atmospheric observers reflected the shifting institutional identities that Agrawal (after Miller & Rose, 1990) associates with *government at a distance* (2005: 193). In order to develop the styles of aggregate governmental knowledge production associated with the regularisation of entire populations, governments have historically developed systems that have enabled them to govern the production of knowledge at a distance. While much is made (as indeed it has been in this volume) of the material objects – such as forms, gauges, optical filters, measurements – that enable government to circulate over extended territorial areas, often overlooked are the new forms of scientific identities that must be nurtured among those who are expected to operationalise the tools of government with science. Crucially, what this volume has shown (particularly through its analysis of early records of smoke observers, the CIAP and the ACAP in Chapters Three, Four and Five) is that the production of scientific subjectivities within agents of atmospheric government has often not been about the subjugation of these individuals to the iron cage of scientific praxis and conduct. Rather records show that it was the local, untrained and underfunded employees of local corporations, and voluntary smoke abatement societies, who sought to be given the capacities to act more scientifically. When given the time, equipment and associated institutional support to act in a more scientific way, it was possible for local air pollution inspectors to see their work and research empowered and spatially mobilised (see Latour, 1988 [1984]). This is precisely why when discussing the relationship between governmentality and subjectivity that care must be taken not to assume that codes of conduct, which are harmonious with the reasons for government, necessarily flow, preformed, from a supposed political centre to a locality. A key feature of governmental power appears to be its ability to harness already-existing subjective desires and visions in order to serve broader schemes of power.

The third, and perhaps most obvious, dimension of the links between atmospheric subjectivities and government uncovered in this book came in Chapter Four, in the form of discussions of the role of clean air exhibitions and new industrial training initiatives in reforming prevailing systems of atmospheric conduct. In many ways the exhibitions and training schemes discussed reveal a more *intimate* expression of the power of government to reshape the everyday conduct of citizens than those associated with the regulations of government at a distance (see Agrawal, 2005: 193–8).[5] They are more intimate to the extent that they served to work new forms of atmospheric concern and conduct into the everyday, not just professional, conscience of citizens. Taking the example of clean air exhibitions first, it is clear that they operated as a distinctly governmental form of technology that was directly focused upon the atmospheric consciousness of the citizen. As governmental technologies, clean air exhibitions were dedicated to presenting subjects with the aggregate impacts a population's mistreatment of the atmosphere could have. From blackened lungs and other biological samples, to the charts and maps that sought to convey the extent of industrial air pollution in Britain, clean air exhibitions attempted to make the atmosphere what Agrawal has termed a 'relevant referential category' around which household and workplace practices could be reconsidered and reformulated (2005: 166). Clean air exhibitions were essentially dedicated to developing an imaginative capacity within the viewer: a capacity that could at once enable them to see like a State sees (i.e. in socio-ecological aggregate) and at the same time better understand their own role in supporting the achievement of governmental goals (i.e. in relation to the creation of newly imagined smokeless homes).

Beyond exhibit pieces, Chapter Four also explored the varied practices of government-sponsored retraining activities that were connected to programmes of atmospheric reform. Some of these retraining initiatives were associated with the clean air exhibitions and sought to support the re-skilling of women in the use of gas and electrical appliances. Beyond the exhibition halls, significant effort was also devoted to the retraining of men working as stokers, boiler attendants and locomotive operators in the more efficient production of combustion. While focusing on very different forms of gender identity, these retraining exercises were connected to the extent that they sought to achieve the governmental goal of atmospheric reform through the production of more scientific modes of subjectivities. Suddenly the woman was no longer a housewife, but a domestic technician – the boiler man a combustion engineer. In his own analysis of environmental subjectivities, Agrawal rightly warns against neatly reading-off changes in subjective actions on the bases of broad categories such as gender, age or race (2005: 172), preferring instead to study closely the impacts of changing modes of governmental subjectivity in their myriad sites of social practice. This volume has not been able to study in depth the impact that new visions of atmospheric woman and manhood, promoted in early twentieth-century Britain, had on atmospheric practices. What analysis has shown, however, is that while not necessarily neatly differentiated according to gender, emerging modes of atmospheric subjectivity promoted within Britain clearly sought to exploit existing gender stereotypes (namely the enslaved housewife and unthinking labourer) as fertile sites to connect atmospheric reforms with broader modes of social emancipation.

The fourth key dimension of subjectivity outlined in this volume focused on the emergence of self-regulating atmospheric subjects. As Chapter Seven illustrated, these new modes of atmospheric self-regulation have been based upon the production and circulation of ever more elaborate, real-time and digital data concerning atmospheric pollution. In essence, this late-modern atmospheric subject has been forged at the intersection of digital environmental knowledge and enhanced computer processing power. Our first encounter with this notion of a self-regulating, digital subject came in relation to discussions of the British government's designation of the *sensitive body*. The sensitive body is a subject who is deemed, on the basis various medical conditions (including asthma and circulatory disorders), to be particularly vulnerable to elevated levels of air pollution. The sensitive body is invoked as a relevant category when the British government's real-time air samples and computer forecasting models detect or predict 'high' or 'very high' air pollution levels. The rapid transfer of this information through various media is meant to offer the sensitive body the chance to change their daily routines and practice in order to militate against the worst effects of the pollution they are likely to encounter. While the designation and protection of so-called sensitive bodies would appear to be a necessary and valuable role of a caring system of atmospheric government, it does raise interesting issues concerning the rationalities of contemporary air policies. The notion of the sensitive body promotes a subject-position of personal care that while asserting the responsibility of government to alert individuals to the risk of impending air pollution events, also partially insulates political authorities from being held accountable for the actual mitigation of harmful levels of pollution in the first place.

The second example of the promotion of more self-regulating forms of atmospheric subjectivity arose in relation to the potential emergence of digital atmospheric beings. The idea of digital atmospheric beings was developed as a concept in this volume in order to explore the new subjectivities that could be produced as a product of the ready availability of online, real-time air pollution data. The operation of the British government's Automated Urban and Rural Network of air monitoring has recently been combined within an online Air Quality Archive to offer new insights into the condition of the British air. While the research presented in this volume did not explore the real impacts that such digital innovations have had on the formation of new atmospheric subjectivities, and associated modes of air conduct, it is clear that such easily accessible data offers the opportunity for atmospheric knowledge to become a more integral part of the social consciousness and individual decision making. Because these new digital technologies of government provide up-to-date and geographically comparable data on air pollution levels throughout Britain, it is possible to imagine the quality of the air becoming a factor in decisions concerning where to buy a home, school your children, or even take your daily run.

It is possible to interpret the contemporary manifestations of a *sensitive body* and *digital atmospheric being* as highly empowering subject-positions. I would argue, however, against an interpretation of the new atmospheric subjectivities that represents them in direct opposition to the more subservient modes of atmospheric selfhood described elsewhere in this volume. While at one level it is clear that the production of atmospheric knowledge that can be readily incorporated into personal everyday decision making is less coercive than the forms of atmospheric persuasion that were exercised in clean air exhibitions and training initiatives, it is also possible to see that the existence of such knowledge forms can shift responsibility for coping with air pollution away from government institutions and place more pressure on atmospheric subjects. These debates about freedom and subjugation aside, I believe that it is crucial to interpret the sensitive body and digital atmospheric being in relation to emerging modes of neo-liberal governmentality (see Larner & Le Heron, 2005). It is analytically important to position these new atmospheric subject-positions in relation to neo-liberalism because they appear to reflect the increasing emphasis that is placed on personal responsibility and self-government within such strands of economic and political rationality. More importantly, however, analyses of neo-liberal governmentalities have revealed a tendency to pursue systems of government that promote socio-ecological care only up to the point that economic gain and profit are not compromised (see Ong, 2005, 2007). It appears that what the contemporary subject-positions associated with the regimes of digital air knowledge and prediction have in common is that they enable economically beneficial levels of uncosted air pollution to continue within a regime that places responsibility for coping with such environmental problems within the rational decision making of the individual citizen.

Learning Like a State in an Age of Climate Change

I have heard historians discuss whether it is possible to really learn from history. Related debates do not, necessarily, question the fact that we can learn about what has happened in the past, but engage with the legitimate issue of whether historical learning can really provide a reliable basis for shaping contemporary decisions and actions. It is clear that the myriad nature of historical reference points and sources mean that the events of the past can, to some extent at least, be weaved together to support almost any lesson that a present generation or political community may want to learn from history. Notwithstanding the naive or cynical manipulation of historical narratives to support contemporary forms of 'social learning', the histories of government and science presented within this volume reveal that while history may not be able to offer paradigms for the present, it can guide our understandings of how to exploit the contingencies of the past in order to support the emergence of different atmospheric futures. Drawing on the conceptual and empirical insights developed throughout this volume, this section reflects on the lessons of air government history in two contexts. First, it considers the parallels that exist between the history of air government and science in Britain and the contemporary apparatus of climate change mitigation that is being developed. Second, analysis concludes by considering the future of air government and science in Britain by exploring alternative regimes of atmospheric knowledge production, which offer more open systems for learning about air and its conduct, and how such systems could be initiated.

Climate change and the era of total atmospheric government

It may appear peculiar that a book devoted to studying the relationship between government, science and air pollution should only begin to discuss questions of climate change in its closing pages. One thing that is now clear (at least within the prevailing scientific consensus) is that the gradual accumulation of human-produced carbon dioxide, and other greenhouse gases, is the single most significant social legacy associated with atmospheric pollution. The reasons why discussions of climate change have not featured more prominently within this volume are threefold. First, although scientific concern for, and awareness of, climate change and the principles of global warming have a long history (dating back to the groundbreaking work of nineteenth-century Swedish chemist Svante Arrhenius; and more recently to the assiduous recordings of changes in the atmospheric concentration of carbon dioxide by Charles David Keeling), a concern with governing climate change has only recently started to occupy the minds of States. It was not until 1984, in the government's response to the *Tenth Report of the Royal Commission on Environmental Pollution*, that we see a firm commitment in Britain to the State-sponsored monitoring of the build-up and effects of carbon dioxide accumulations in the atmosphere (DECDEP, 1984: 71–2). Second, the fact that carbon dioxide is an abundant compound in the Earth's atmosphere (a fact that famously, if erroneously, led to the claim that carbon dioxide was not an air pollutant at all), whose polluting effects are only expressed at the level of global heat balances, means that its study has not required the same forms of spatial orchestration between government and science that has been seen with other forms of air pollution. In a sense the study of carbon dioxide has moved in completely the opposite direction from the longer historical study of other air pollutants. Whereas the study of carbon dioxide has moved from the compilation of aggregate global assessments of atmospheric concentrations, to a retrospective analysis of the national and local sources of such pollution, the historical knowledge of aerosol and other air pollutants has gradually accumulated from local assessments of air quality to provide much larger geographical pictures of the extent and movement of pollution. Thirdly, and finally, it is clear that the assessment and government of carbon dioxide in Britain has largely been conducted within the frameworks of air pollution assessment and control that had already emerged in order to regulate other air pollutants.

Notwithstanding the reasons why climate change has not featured more prominently within this volume, it is clear that there are some crucial parallels between the history of air pollution presented here and contemporary governmental and scientific practices surrounding climate change. At one level, despite being a very different form of pollutant than those already discussed, the gradual accumulation of carbon dioxide in the Earth's atmosphere has been part and parcel of the same socio-industrial processes that have created the atmospheric problems discussed throughout this book. It is helpful to think of carbon dioxide as the invisible and insidious pollutant that escaped the watchful gaze of the atmospheric observers; was undetected in the gauges of the CIAP; and was untested by the filters of the National Air Pollution Survey, to only now become an object of government with science. But the parallels run even deeper than this. It is interesting for example to think of climate change as the meteorological apex of air pollution history. We saw that the early monitoring of air pollution in the late nineteenth and early twentieth centuries was tightly tied to the expertise of meteorological experts. While the early science of air pollution was, however, concerned with the role of the weather in conditioning the severity and location of pollution events, it appears that with the advent of climate change the government of air pollutants is being used as a way of controlling aspects of the climate and meteorology. There is, of course, another rather ironic connection between climate change and the forms of air pollution analysed in this volume. There is a concern that aerosol pollution (including smoke and sulphates) may actually be shielding certain parts of the planet (particularly heavily industrialising metropolitan areas) from the worst impacts of global warming (see Pearce, 2007: 177–83). While scientists remain uncertain about whether aerosol pollution is actually contributing towards or militating against global warming, it is possible that if the fight for cleaner air is finally won, it could release up to a quarter of existing atmospheric warming, which is currently being held up by dirty air, into the global climate system (ibid.: 179).

Although it is clear that the climate change debate and the history of air pollution government presented in this volume reflect parallel and, at times, directly overlapping stories, it is important to consider in greater detail precisely what additional perspectives our discussion of atmospheric government with science can have for the contemporary deliberations concerning climate change and associated political and scientific practices. At one level, it is clear that contemporary debates around the nature and regulation of climate change are recreating many of the tensions between States and atmospheric science we have charted throughout the history of air government in Britain, but at an increasingly international level (see Miller, 2001; Miller & Edwards, 2001a). While the British government could now be classified as a strong advocate of climate change abatement, it is clear that many more sceptical political administrations are questioning the nature of climate change on the basis of the science upon which its discovery has been based. It appears that when it is politically expedient for government authorities to either question, or outright undermine, arguments for anthropogenic forms of climate change, it is the work of atmospheric scientists that has come under the most intense scrutiny. It is in this context that discourses of *junk science* have been used to question the rigour and validity of the work of atmospheric sciences (see Monbiot, 2006: 20–42).[6] It is not just that government authorities have been deliberately undermining and ignoring climate scientists – although there are times when they clearly have. There is genuine uncertainty in the scientific community (within a prevailing consensus on the nature and causes of climate change) as to the precise extent of climate fluctuations we may be facing, and what the best course of action should be. What is clear is that the concerns over the role of the 'speculative' sciences of air pollution we have seen throughout the history of atmospheric government is going to be a recurring theme within emerging regimes of climate governance.

The nature of climate change means that its effective study requires the formation of unprecedented programmes of international scientific collaboration. It is in this context that the twenty-first century is being characterised by new fusions of atmospheric science (in particular climate sciences and modelling) and international government systems (see Miller & Edwards, 2001b). These programmes of collaboration do not only need to cross national borders, but also require the careful calibration of research conducted on glaciers, oceans, vegetation mass, land-use change, the carbon cycle, global albedo levels, as well as climatology and meteorology. While such research is currently being supported and marshalled (if not directly coordinated) by key United Nations organisations and the Intergovernmental Panel on Climate Change, it is clear that such a total science of the climate is going to provide the basis for emerging forms of atmospheric government with science in the future. It is precisely in this context that much research needs to be done on the emerging relations between governments and the inevitably speculative sciences of climate change, in order to better understand how government systems can foster effective apparatus of climatic knowledge gathering, and climate science can support nascent systems of environmental government at a range of scales.

Beyond the changing relations between atmospheric science and government that are connected to the climate change debate, it is interesting to note the strong relations that exist between current regimes of carbon dioxide monitoring and systems of air pollution surveillance. While British greenhouse gas emissions are not monitored systematically through a physical network of automated and non-automated sampler sites,[7] the British government does deploy the same system of surrogate statistical analyses as used to monitor other forms of air pollution in order to compile greenhouse gas statistics. Consequently, British greenhouse gas emissions are compiled by AEA Energy and Environment from existing statistics concerning energy consumption, transport, industrial production, agricultural activity and changes in land use and forestry (the process for calculating the methane released from landfill sites is currently being reviewed by AEA Energy and Environment).[8] It is on the basis of these statistical sciences that it is possible to estimate that the UK emits approximately 468 million tonnes of carbon emission every year.[9] It is important to recognise that while the statistical estimates of greenhouse gases in Britain appears impressive in its scope it is actually highly limited. At present there are key sectors of the economy that are not incorporated in the measures (including most controversially the aviation industry). Also, current greenhouse calculations do not account for so-called *embedded emissions* – the emissions that are produced outside the UK by British companies that have relocated *inter alia*.

It is on the basis of such estimate techniques that, as with other air pollutants, the British government assesses its levels of exceedence of greenhouse gas emissions. It is at this point that emerging sciences of climate change modelling and prediction are starting to undermine regimented regimes of atmospheric government. Currently Britain is committed to a United Nations Framework Convention on Climate Change agreement to reduce its greenhouse gas emissions by 12.5% below 1990 by 2008–12, and a new domestic goal of a 60% reduction in carbon dioxide emissions by 2050.[10] As we saw with the setting of aerosol-based forms of air pollution exceedence levels, current scientific research cannot provide absolute certainty concerning the levels at which air pollution will cause long-term harm to people and the environment. Such concerns with the setting of reliable thresholds for atmospheric government do, however, appear to becoming an even greater concern within climate research. Increasingly scientists are coming to realise that complex changes in the environmental systems that regulate, and are in turn regulated by, the atmosphere (namely the cryosphere, oceans and biosphere) are unlikely to follow liner or incremental patterns of transformation (so-called *Type I* changes) (see IPCC, 2007; Pearce, 2007). Instead it appears possible that climate change could create a series of nonlinear, sudden transformations (so-called *Type II* changes) in global atmospheric conditions that make current greenhouse emissions targets and the time-scales over which they are set problematic.[11]

The final parallel concerns the types of atmospheric subjects that are being promoted in association with emerging systems of climate government in Britain. It is clear that with the incorporation of climate change learning within the national educational curricula that there is a pedagogic State strategy to create a new atmospheric consciousness within the British citizenry. While paralleling the forms of direct instruction that surrounded early strategies of pollution abatement in Britain, there is an argument that with the effects of climate change often being distantiated from the individual subject (both in time and space) that creating an imaginative link between the citizen and the atmosphere is now more important then ever. Beyond the collective provision of climate change education, it is also apparent that systems of climate government are attempting to instil self-reflective structures for atmospheric reform. These strategies of atmospheric self-reflection have been targeted at individuals through the government's *Act on CO₂* initiative, and at corporations and institutions of various kinds thorough the work of the *Carbon Trust*. These initiatives build on the current fashion for personalised carbon management plans, calculators and gyms in order to enlist the individual, or organisation, in the active monitoring and assessment of their own atmospheric relations. These strategies seek to cultivate senses of self-interest in carbon management (particularly in relation to the financial savings associated with energy conservation). But, unlike the forms of subjectivity associated with air pollution abatement we have discussed, carbon calculation appears to make the climatic self in a different mould than the atmospheric subject. Carbon calculation actually engages the subject in the process of (virtual) air monitoring. If Agrawal is correct, and personal engagement in the processes of environmental monitoring promotes a much deeper sense of engagement and care for the environment, then what is occurring in carbon management could offer the hope of different types of atmospheric self. This could be a self who is more intimately engaged in the processes of atmospheric knowledge production, not merely a subject who is charged with responsibility for insulating themselves against the worst side effects of air pollution.

Cartographies of the atmospheric future: on collective learning and new air mentalities

At the beginning of his 1979 lecture course at the Collège de France (translated as the *Birth of Biopolitics*), Foucault reflects on the intentions that informed the previous year's lectures on the history of governmental reason (Foucault, 2008 [2004]). Crucially, Foucault muses that his intention in constructing a history of government had not been to undermine the vision of governmental institutions as a potential power for good, but to unstitch the historical universals of State, society and economy in order to reveal the possibility of other reasons and systems for government (ibid.: 1–4). This volume has neither been a celebration, nor a denigration of the history of air pollution government in Britain. Instead it has sought to unpack the historical universals of State and science in order to reveal how systems of atmospheric government with science have been constructed around the contingent opportunities for the production of atmospheric knowledge. What has become clear is that the seamless, and seemingly overwhelming, power of contemporary knowledge systems concerning the atmosphere tend to reinforce the belief that only particular constellations of big State and big science could possibly comprehend the complexities of the air and guide associated forms of government. The uncontested supremacy of atmospheric government with science in the production of knowledge is a basis for a largely unchallenged power over the air. But it is worth reflecting on this axiomatic connection between power and knowledge that is so often made within political analysis. While it is clear that knowledge is a crucial context for the securing of power, it is also clear that not all power is knowledge-based, and that not all knowledge necessarily becomes powerful. In this context, what is most striking about the history of British air pollution science and government is how little of the things we know about the fluctuations of the air have actually become the basis for fundamental changes in the way in which we relate to the atmosphere. Acknowledging the effectiveness of the 1956 Clean Air Act (and associated National Air Pollution Survey) in overcoming Britain's 'smoke problem', most atmospheric knowledge has had little lasting impact on prevailing social patterns of behaviour concerning air pollution. While we may look back in wonder at the smogs of 1950s' Britain (or the contemporary air quality of Beijing) and shake our heads in disbelief, we collectively continue to pollute the atmosphere through the invisible emissions of our car exhausts and embedded carbon footprints. The power of atmospheric knowledge consequently appears to have predominantly been expressed as a basis for justifying tolerable levels of pollution and redefining basic rights to pollute. My concluding question then becomes: how can we make atmospheric knowledge a more powerful force for socio-political change?

In considering this question I started to reflect upon why the vast collections of atmospheric knowledge produced in Britain every day appear to have only intermittent relevance for contemporary forms of air conduct and politics. It appears to me that the reasons for this disjuncture can be discerned within the contemporary structure of atmospheric knowledge production in Britain. With the possible exception of personal carbon dioxide calculations, the vast majority of atmospheric data on air pollution compiled in Britain is done so either as part of nationally orchestrated government networks, or specialist scientific research programmes. The atmospheric knowledge produced as part of these large-scale systems tends to emanate from spaces and places that have been chosen for their statistical and/or scientific utility, but rarely tend to have meaningful connections to social territories. Under this system, atmospheric knowledge only becomes meaningful to subjects when it is abstracted from its points of production in networks of government science to constitute a regulatory basis for personal best conduct. While it may be difficult to imagine legitimate alternatives to current systems of atmospheric knowledge gathering, history does provide us with some clues. One thing that the histories presented in this volume have shown is that far from being the reserve of a national elite of government bureaucrats and designated atmospheric scientists, the monitoring of air pollution emerged out of local networks initially forged by concerned citizens and scientists. These citizen-scientists used their knowledge of local areas and emerging scientific techniques as a way of calibrating atmospheric surveillance with visions of community renewal and improvement. While I do not wish to romanticise the air pollution scientists of the past, the local subjectivities of previous collective experiments in atmospheric knowledge production could offer inspiration for the twenty-first century. It is interesting to imagine the impact that rescaled systems of atmospheric surveillance, which while still replete with the modern accoutrement of atmospheric sciences could be constructed around spaces of collective meaning and identification, like the community, neighbour or turf. It is not that these new atmospheric spaces would be disconnected from other urban, regional, national and global networks of knowledge gathering, but they would allow new forms of collective engagement with the atmosphere. If we take Rose's call for cartographies of the future literally, is it not possible to imagine the spatial reorganisation of atmospheric surveillance on these terms providing the basis for a new set of mentalities towards the air, and novel practices of air government?

Inspiration for such cartographies of the future does not, however, lie only in the past. At the moment there are a number of examples of collective forms of air monitoring being developed and implemented. Certain local authorities in Britain, for example, are encouraging the online compilation of community carbon footprints, which are enabling local areas to understand their local relations with the atmosphere.[12] In the USA a number of community air monitoring programmes have already be initiated. A prominent example is the Environmental Protection Agency's *AirBeat* initiative that operates in the Roxbury Neighbourhood in Boston, Massachusetts (see Environmental Protection Agency, 2002).[13] This programme, which is based upon an online real-time pollution analysing device, appears to be very similar to the automated urban and rural networks of air monitoring that operate in Britain. However, through strong links with local environmental groups and research institutions, which helped set up and locate the monitoring equipment, the AirBeat programme has been able to integrate air monitoring into a broader politics of urban air defence and community engagement. A number of community groups have consequently used the data produced by the scheme to support campaigns for clean buses and environmental education drives (ibid.). Such schemes are important because they appear to move beyond individualised monitoring to engage in a form of collective learning. Collective learning, constituted within meaningful spatial contexts, appears to be empowering in ways which more individualistic visions of atmospheric subjectivity and government are not. Collective atmospheric learning can help communities to understand how a number of issues ranging from new housing, road and industrial developments, to the relocation of school and healthcare facilities, can change a community's air and people's relationship to it. To paraphrase Latour, 'matters of atmospheric fact suddenly become matters of political concern' (Latour, 2007a: 5). Ultimately such initiatives are empowering to the extent that they shift atmospheric politics away from the normative realms of personal reform and into the arenas of political geography. It also appears likely that the construction of community atmospheres will provide a more powerful context within which to lobby for governmental change and air policy reformulation than those offered within more individualistic forms of air subjectivity.

While I believe studies of the new atmospheric spaces of collective science could provide a fertile terrain for much research in human and physical geography over the next decade, this call for alternative ways of knowing and governing the atmosphere should not be interpreted as a neo-anarchist attack on the State. States, as key nodes and progenitors of atmospheric government with science, have a crucial role to play in our collective air futures. As Latour presciently reflects, 'How can we detect new phenomena at the extreme limit of the sensitivity of instruments, without a meticulous accumulation of data over a very long time? No one has the ability to keep track of these except administrators' (2004: 205). In this context, I am with Latour when he recognises that it is not States that disenfranchise us from atmospheric power and responsibility, it is the ideologies of scientific and governmental elitism, and associated obfuscating modes of atmospheric calculation that accomplish this task (2007b). Rather than placing our faith in the cognitive rationalities of States to think through our atmospheric problems for us then – which ultimately appears to lead to governmentalities that offer minimal environmental protections and only guard narrow thresholds of ecological security – we need to find new ways of *learning with States*.[14] We need more, not less, air sciences to support more, not less, atmospheric government.

Notes

1 For more on this most probably apocryphal tale see Brimblecombe (1987: 9–14).
2 While rightly suspicious of accounts that suggest an early transgression of the 1306 proclamation was decapitation, Brimblecombe does propose that the destruction of furnaces was a likely penalty for disobeying the Proclamation as this was the established punishment for those building furnaces on roadsides (1987: 9). It is very difficult to locate any reference to this much-reported exercise of torture that cites an original historical source as evidence for its occurrence.
3 See the *British Air Quality Archive* at http://www.airquality.co.uk/archive/index.php (accessed 8 February 2008). The nature and work of the Air Quality Archive will be discussed in far greater detail later in this volume (see in particular Chapter Seven). The Archive is supported by the Department for Environment, Food and Rural Affairs, the Scottish Government, Welsh Assembly Government, and the Environment and Heritage Service, and is run by AEA Energy and Environment.
4 For a detailed analysis of the history and operation of the 1843 Select Committee on Smoke Prevention see Ashby and Anderson (1981: 7–11); Brimblecombe (1987: 101–3); Mosley (2001: 119–20).
5 For an informative overview of the air pollution levels in Beijing and various attempts that are being made to measure and government particulate matter see British Broadcasting Corporation (2008).
6 See World Health Organisation (2005: 9–14) for more information on air quality standards (and the rationality behind such standards). Note the level of 50 micrograms/cubic metre pertains to permissible levels of PM_{10} for a 24-hour mean.
7 These figures are based on readings taken at the Olympic Village and BBC office in Beijing by the British Broadcasting Corporation (Bristow, 2008). Official air pollution figures can be obtained from the Beijing Municipal Protection Bureau at http://www.bjepb.gov.cn/air2008/olympic.aspx (accessed 11 August 2008).

8   These figures were based on estimates made in 1995/96 (Royal Commission on Environmental Pollution, 2007: 35). For more information on links between air pollution and human health in Britain consult reports produced by the Department of Health's Committee on the Medical Effects of Air Pollutants (COMEAP).

9   This estimate was actually based on figures derived from Defra (2006).

10 I would like to acknowledge at this point the contributors to a session I co-convened (with Simon Naylor) at the 2007 Annual International Conference of the Royal Geographical Society (with the Institute of British Geographers) in London. The session was entitled *Atmospheric Geographies: The Politics and Histories of the Skies*, and brought together human and physical geographers and historians of science. This session, and the sense of supportive collaboration it embodied, has been a great source of inspiration for me in the completion of this volume.

CHAPTER TWO

1   The degree of Molesworth's respect and admiration for 'men of science' is easy to discern from his inaugural address to the *Manchester Association for the Prevention of Smoke* (see Mosley, 2001: 119). In this inaugural meeting Molesworth described how he felt like a 'dwarf amongst giants' – 'a child amongst sages' when in the company of men of science (ibid.: 119).

2   Ibid.: 7.

3   See for example Ashby and Anderson's (1981) account of the early work of Michael Angelo Taylor MP to establish a Parliamentary Act to abate urban smoke in Britain, ibid.: 1–7.

4   HC.PP.1843(583) – Final Report: iii.

5   In London the *Assize of Nuisance* was used as a mechanism for resolving air pollution disputes between metropolitan neighbours (see Brimblecombe, 1987: 12–14).

6   *Fumifugium; or the inconvenience of the aer and smoak of London dissipated* was first printed by W. Gobdin for Gabriel Bedel and Thomas Collins, London. An online edition of *Fumifugium* is available at http://www.geocities.com/Paris/LeftBank/1914/fumifug.html (accessed 8 August 2007).

7   In his wide-ranging analysis of *Fumifugium* Mark Jenner (1995) argues that Evelyn's work should not be interpreted simply as an act of environmental benevolence, but as a highly political act. Jenner discerns great significance in Evelyn's dedication of *Fumifugium* to King Charles II (1995: 537). *Fumifugium* was published the year after King Charles II's restoration to the throne, and Jenner argues that it was the desire to usher in a new political era, as much as a moral opposition to the evils of air pollution, that infused Evelyn's work (cf. Brimblecombe, 1987: 47–52). As a devoted Royalist, Jenner argues that Evelyn used *Fumifugium* to fuse political ideology and ambition with air pollution. To these ends, the smoky chaos of London became synonymous with the interregnum, while the desire to produce a clean and healthy atmosphere in London becomes a symbol of the inherent virtue of the new political regime (ibid.: 540).

8   In order to fulfil his smokeless visions, Evelyn proposed the use of alternative fuel sources to coal, the movement of smoky trades to the outskirts of the metropolis and creation of new parks with flowers that would perfume the atmosphere (see Brimblecombe, 1987: 50; Jenner, 1995: 544).

9   Brimblecombe does suggest, however, that it may well have been the costs of Evelyn's plans that meant it was unlikely to receive Parliamentary support and was thus prohibitive to implement on a large scale (ibid.: 50).

10 HC.PP.1843(583)-Final Report: 177–80.

11 HC.PP.1843(583)-Final Report: 180. In this quote Faraday was making specific reference to the practicability of smoke abatement in the domestic sphere.

12 It is interesting to note that while at the Royal Manchester Institute Lyon Playfair worked with Angus Smith, who would later head the Alkali Inspectorate and coin the term acid rain.

13 HC.PP.1846 (194). De la Beche and Playfair were actually instructed to pay particular attention in their study to the towns of Leeds, Manchester, Bradford and Derby.

14 HC.PP.1846 (194): 3–4.

15 HC.PP.1846 (194)-App D. This appendix provides details of the observations of smoke made by Joseph Fox at the Cotton Mill on David Street in Manchester.

16 It should, perhaps, not be surprising that Foucault rejected his own scientific characterisation of modern systems of government. In his earlier analysis of the *Archaeology of Knowledge* Foucault displayed a keen awareness of the distinctions between scientific epistemologies, with their 'coherence and demonstrativity', and *quasi-sciences* (such as political economy) (Foucault, 2006 [1969]).

17 As perhaps the first widely applied European philosophy of economic government, the Physiocratic system claimed that a nation's wealth and well-being were based upon the effective use of productive agricultural lands. The Physiocratic position marked a strong contrast with competing *Mercantilist* philosophies that associated wealth with the accumulation of financial resources in the ruling elites, see Charbit (2002).

18 This point can, in fact, be discerned in his earlier lecture to the State University of Rio de Janeiro in 1974. In this lecture he introduces the idea of a *science of the state*. According to Foucault the notion of a science of state could be seen in two main ways: (i) as an arena of research which focuses on the State itself as an object of analysis (something akin to political science); or (ii) as a reference to procedures in and through which State bureaucracies collect and collate knowledge that constitutes the basis for governmental decision making (Foucault, 2000a [1994]).

19 See Popper, 2002 [1950]: 27–34 and his discussion of science as methodology.

20 For a more detailed review of the epistemological upheavals associated with the scientific revolution see Shapin and Schaffer, 1985; Shapin, 1988, 1994, 1996.

21 Here Foucault is quoting Chemnitz's *Dissertatio* volume 1 (1712 [1647]): 6.

22 The uncertainties surrounding the nature of scientific method are articulated clearly in the foundational debate between Rudolf Carnap and Karl Popper concerning whether scientific methods and data are defined by processes of

verification or falsification. For a short overview of this extensive and extended debate see Hacking, 2005 [1983]: 1–17.

23 It is precisely in this context that statistics, as a key modality of standardised, scientific measurement (and quite literally meaning, of course, 'the science of the state'), has been interpreted as a key indicator of the rise of the governmental State (see Hacking, 1990).

24 For a very interesting discussion of what being anti-scientific may mean (at least in relation to the contemporary *science wars*) see Shapin, 2001: 99–115.

25 See Popper (2002 [1950]: 3–26) for a discussion of how belief in the certainties of science, as oppose to its ever-changing and evolving method, has been exploited to justify the unchanging logics of authoritarian and totalitarian regimes. In these instances, while purportedly based on scientific logic, such State systems actual undermine the methodological premise of scientific discovery.

26 Foucault, 2007 [2004]: 358.

27 It should come as little surprise that Foucault commenced his 1978 lecture series with an exegesis on biopower because this is the concept upon which he terminated his previous lecture series at the Collegè de France, (published in English as *Society Must Be Defended*) in 1976 (Foucault was on sabbatical in 1977) (Foucault, 2004 [1997]).

28 Foucault, 1991 [1975], 2002 [1961], 2003a [1963].

29 While it is possible to trace Foucault's interest in biopower to his earlier excavation of the techniques of bodily observation and discipline associated with prison, clinic and asylum, it is not until 1974 that we see an explicit problematisation of the term, see Foucault, 2000a [1994]. See also Foucault, 1998 [1976]: 135–59, 2004 [1997]: 239–64.

30 The emerging role of governmental institutions within the *administration of life* can be seen in three broad ways. First, and at a more local level, the role of government authorities in the management of health can be discerned in the rise of public health and the associated regulation of the urban hygiene, housing and sanitation that commenced in eighteenth- and nineteenth-century Europe (what Foucault describes as *urban medicine*) (Foucault, 2000a [1994]: 134). Second, from the eighteenth century onwards it is possible to see the increasingly active State control, standardisation and certification of the medical profession (what Foucault termed *state medicine*) (ibid.). Thirdly, and supported by the establishment of a State-sponsored medical cadre, the governmentalisation of the medical agenda was predicted upon the collation of national figures concerning the health of the population (including statistics of fertility, morbidity and longevity).

31 It is in this context that contemporary writers such as Agamben and Dillon have emphasised the complex systems of co-existence between sovereignty, discipline and security and sought to uncover the *dark side of biopolitics*, see Agamben, 1998; Dillon, 2004.

32 It is clear from the work of Foucault that although the rise of biopower to the level of demographic calculation is connected to the rise of a governmental state, that governmental rationality has its own history that cannot be circumscribed to biopolitical rationality alone.

33   For an overview of the range of meanings associated with governing in the thirteenth, fourteenth and fifteenth centuries, for example, see Foucault, 2007 [2004]: 121.

34   Drawing on the reflections of Saint Gregory Nazianzen, Foucault ultimately claims that this history of *governing men* is essentially a history of the 'art of arts' and the 'science of sciences' (Foucault, 2007 [2004]: 150–1).

35   Here Foucault considers the various meanings associated with the verb to govern, According to Foucault one of the key historical definitions of the verb to govern is 'to conduct someone'. See also Foucault, 2007 [2004]: 193, where he outlines the different ways in which it is possible to understand the notion of conduct.

36   For an excellent overview of the diverse methods and epistemologes of SSK and its relations with STS and the history of science see Shapin, 1995.

37   Haraway famously called for a situated account of scientific knowledge as an epistemological and methodological remedy for what she discerned to be a widely accepted vision of an all-seeing science that floats above the reality it reveals (Haraway, 1991: Chapter 9).

38   See Foucault's extended discussion of the relationship between subjugated knowledges and science in Foucault, 2004 [1997]: 6–14.

39   Note that Latour positions his analysis of pasteurisation within the epic account of the Napoleon's 'Russian Campaign' provided by Leon Tolstoy's *War and Peace*. According to Latour, just as Tolstoy's novel reveals that the complexities of military battles make it difficult to ascribe victory or defeat to the actions of often-isolated leaders (whether it be Napoleon or Kutuzov), the success or failures of scientific techniques are never the product of the actions of scientists alone (Latour, 1988 [1984]: 3–12).

40   This form of methodological perspective is precisely why Latour argues that contemporary sociology must abandon its assumption that 'the social' (whether it be in terms of the sociology of scientific knowledge, or the social construction of nature) is a category of explanatory power. Latour argues that it is society that must be reassembled and explained. For an earlier rendition of this argument see also Latour, 1988 [1984]: 9. For a related discussion of the problematic deployment of nature within science and politics see Latour, 2004: 9–52.

41   For more on the nature and history of the *sciences war* see Shapin, 2001. According to Shapin the science wars are based upon the fact that claims about the social nature of scientific knowledge have been asserted so strongly by sociologists rather scientists themselves not recognising the contingent nature of scientific enquiry.

42   See Michel Senellart's essay on the context of Foucault's 1978 lecture series in Foucault, 2007 [2004]: 369–401 and his discussion of Foucault's relationship with French politics and his search for a socialist govermentality that could help to reinvent the Left.

CHAPTER THREE

1   Smoke Consumption Report, 31 January 1925, L.Met.Arch LCC/PC/Gen/1/9.

2   L.Met.Arch LCC/PC/Gen/1/9.

3 See, for example, the records of Birmingham's borough analyst Alfred Hill on the detection of acid-based forms of air pollution by use of olfactory classifications, B.Cit.Arch BCC/AR-76.

4 See in particular the evidence given to the 1843 Select Committee on Smoke Prevention by Captain A.W. Sleigh (Assistant Commissioner of Police in Manchester), presented on 3 August 1843. Under questioning by Committee members Captain Sleigh recounts his visual survey of chimneys in Manchester and his assessment of the relative contribution of domestic and industrial premises to the air pollution problems of the city. Within the records of this testimony it is clear that members of the 1843 Committee remain unconvinced with the ability of Captain Sleigh to effectively differentiate between the quantities of smoke emanating from industrial and domestic fires. HC.PP.1843(583) – Final Report para. 1521-1561.

5 Crary utilises Deleuze and Guattari's notion of *amalgamation* to reposition vision as the historical product of a range of material and discursive practices. The notion of amalgamation is developed in Deleuze and Guittari (1987) to reveal how the power of technology is not the product of the tool itself, but the relations between the tool and social context.

6 For more on the complex nature of modernity as 'vital mode of existence' see Berman (1983).

7 It is important to note that the necessary (re)embracing of subjective vision during the nineteenth century did not involve the abandonment of objective perspective. The work of Donna Haraway reveals how the ideal of the *camera obscura* (the insulated site for enhanced sight) has continued to inform the scientific ideology of 'seeing everything from nowhere' and been used to deny the situated nature of all vision (Haraway, 1991: 183–201).

8 Crary describes the transformation of the science of sight in the nineteenth century as a movement from the abstract geometric optics of classical science to the physiological optics associated with the corporeal experience of modernity (1992: 16).

9 For an interesting discussion of the antagonistic relationship between Foucault's analysis of vision, discipline and surveillance, and Guy Debord's theory of the *Society of Spectacle* see Crary (1992: 17–19).

10 Note here the similarities in technique between nineteenth-century scientific sight and what Urry terms the *romantic gaze*. As with trained scientific sight, the romantic gaze of the eighteenth and nineteenth centuries was synonymous with solitude and intense reflection. However, while the scientific gaze used time to engage reason, the romantic gaze sought to cultivate a sense of the irrational as a basis for appreciating the natural world (see Urry, 2002: 43–5).

11 It was in fact in 1848, as part of a Public Health Bill, that the British State first offered a national law prohibiting certain forms of air pollution. One of the key consequences of the 1848 Act was the formation of a National Board of Health and corresponding Local Health Boards to implement the legislation and monitor public health.

12 B.Cit.Arch BCC/AR-74.

13 B.Cit.Arch BCC/AR-74.

14 B.Cit.Arch BCC/AR-74.

15 B.Cit.Arch BCC/AR-74; B.Cit.Arch BCC/AR-76; B.Cit.Arch BCC/AR-78.
16 B.Cit.Arch BCC/AR-74; B.Cit.Arch BCC/AR-76; B.Cit.Arch BCC/AR-78.
17 B.Cit.Arch BCC/AR-76.
18 For further details of this arrangement and the circumstances surrounding it see B.Cit.Arch BCC/AR-78.
19 According to Mosley the Manchester Police Commission was given significant responsibility for the monitoring and regulation of air pollution offences from the late eighteenth and early nineteenth centuries.
20 B.Cit.Arch BCC/AR-73, Minute 5437 – 7 February 1877.
21 The Alkali Acts were a product of an 1862 Parliamentary Select Committee chaired by Lord Derby. Thorsheim claims that the Alkali Act of 1863 resulted in the creation of the first nationally scaled body for regulating environmental pollution in the world (Thorsheim, 2006: 114).
22 For more on the role of factory inspectors and the remit of their respective spatial responsibilities see Jones, 2007: 111–42.
23 Report to Birmingham Inspection Committee dated 13 June 1866 – B.Cit. Arch BCC/AR-76.
24 The technological fixes that were available for the absorption of acidic air meant that the Alkali Acts were successful in reducing the emission of hydrochloric acid by greater than 95% in a relatively short period of time, see Brimblecombe, 1987: 139.
25 Birmingham Borough Inspection Committee (1872). *Report on the Public Health 1872, the Steam Whistles Act, and the Adulteration of Food etc., Act for Presentation to the Council* (Steam Printing Offices, Birmingham).
26 Brimblecombe parallels the emergence of Sanitary Authorities with more specialist training regimes and systems of accreditation. In 1876, for example, the Sanitary Institute created a system whereby sanitary inspectors could be trained and certified (Brimblecombe, 2004: 17).
27 For more on the operation of the Smoke Sub-Committee see B.Cit.Arch BCC/AR-72.
28 B.Cit.Arch BCC/AR-78, Minute 3738.
29 B.Cit.Arch BCC/AR-78, Minute 3756.
30 A range of para- and nongovernmental agencies now produce statistics, but the etymology of the word (from the German *statistik*) literally translates as 'knowledge of the state' (Foucault, 2004 [1997]: 274).
31 B.Cit.Arch BCC/AR-73 – 4 January 1877.
32 B.Cit.Arch BCC/AR-73, Minute 5437 – 7 February 1877.
33 In September of 1876 alone 756 observations of smoke pollution were recorded in Birmingham, B.Cit.Arch BCC/AR-73.
34 HC.PP.1843(583)-App.6, pages 202–208.
35 HC.PP.1843(583)-App.6, page 206. See also Mosley (2001: 139). Interestingly, George Orwell recognised the role of smoke in abetting the activities of polluters at a much later point in time in his famous account of working-class Britain, *The Road to Wigan Pier* (1937). Describing one of his 'urban rides' into Sheffield Orwell writes of his failed attempt to simply count the numerous chimneys, '[o]nce I halted in the street and counted the factory chimneys I could see; there were thirty-three of them, but there would have been far more if the air

had not been obscured by smoke' (Orwell, 1937: 35). There appears to be a particular irony in the way in which smoke as an observable indicator of air pollution tends to work against its own regulatory observation.

36 B.Cit.Arch BCC/AR-72; B.Cit.Arch BCC/AR-73.

37 A copy of the 1875 report produced by Birmingham's Sanitary Committee is available in the minute book B.Cit.Arch BCC/AR-72.

38 B.Cit.Arch BCC/AR-72, 16 February 1875.

39 *Report of the* [Birmingham] *Sanitary Committee for Presentation at a Special Meeting of the Council*, 27 July 1875 – B.Cit.Arch BCC/AR-72.

40 B.Cit.Arch BCC/AR-73, Minute 5401 – 3 January 1877.

41 L.Met.Arch LCC/PC/Gen/1/33.

42 L.Met.Arch LCC/PC/Gen/1/33.

43 As Golinski (2006) has revealed, a concern with the impacts of the varied qualities of the air (including its moisture content, pressure and particulate content) on British health and social sensibilities had defined areas of overlapping scientific interest for meteorologists and pollution experts as early as the seventeenth century.

44 L.Met.Arch LCC/PC/Gen/1/33 – Smoke Observation Report, 28 April 1899.

CHAPTER FOUR

1 See Peter Thorsheim's discussion of the work of the Fog and Smoke Committee of the National Health Society (Thorsheim, 2006: 91–9).

2 Morus's *Frankenstein's Children* (1998) provides a fascinating insight into novel combinations of experimental science, entertainment and exhibition in the promotion of electricity in early-nineteenth-century Britain. The promotion of electrical power, along with other 'smokeless technology', would, of course, become a crucial part of the clean air exhibitions of the late-nineteenth-century Britain.

3 It is for these reasons that when analysing manifestations of cultural power Bennett turns to Antonio Gramsci instead of Foucault.

4 Note again here the intellectual antagonism mentioned in the previous chapter that existed between Foucault's account of the disciplinary society captured in *Discipline and Punish* and the vision of power in a more consumer-based society developed in Guy Debord's *Society of Spectacle* (cf. Foucault, 1991 [1975]; Debord, 1992).

5 See Thorsheim's analysis of the networks of reform that supported the formation of clean air exhibitions (2006: 88–91).

6 Thorsheim draws particular attention to the role of the *Social Science Association* and its organisation of sanitary reform exhibitions in the 1870s as a key stimulus for the clean air exhibitions that emerged in the 1880s (2006: 80–1).

7 The Smoke Abatement Committee became the Smoke Abatement Institute following its incorporation, which was supported by the Board of Trade (*The Times*, 1882).

8 Ibid.

9 For a more detailed history of the Smoke Abatement Committee and the role of Ernest Hart and Octavia Hill within it see Thorsheim (2006: 88–99).

10 See Anderson's (2003: 422–41) discussion of the use of competition as a basis for agricultural reform and improvement in Australia for example.

11 County Municipal Record (1912) 24 September: 510, L.Met.Arch LCC/PC/ Gen/1/25.

12 *Glasgow Herald* (1910) 'Smoke abatement, gas fuel, appliances at the exhibition', 11 September, G.Cit.Arch MP40/208.

13 *Glasgow Herald* (1912) 'Smoke abatement in Glasgow: exhibition opened', 21 September, G.Cit.Arch MP40/208.

14 Sir Oliver Lodge was a vice president of the National Smoke Abatement Society.

15 A broader set of spatial plans and associated photographs of the layout of the smoke abatement and clean air exhibitions of the National Smoke Abatement Society is available at HLG/55/209.

16 Among those organisations who loaned material to the 1936 exhibition were: His Majesty's Office of Works; the Corporation of Manchester's Public Health Department; the Royal Botanic Gardens, Kew; the British Leather Manufactures Research Association; and the government's Chemical Research Laboratory. These exhibit pieces were joined by material donated by a number of individual scientists and photographers.

17 The centrepiece of the Leather Manufactures Research Association's exhibits was a leather-bound book from Aberystwyth. While several centuries old, it was claimed that this volume was in much better condition than recently bound volumes stored in industrial towns (National Smoke Abatement Society, 1936: 49).

18 L.Met.Arch LCC/PC/Gen/1/25.

19 G.Cit.Arch MP40/208; L.Met.Arch LCC/PC/Gen/1/33.

20 *Glasgow Herald* (1912) L.Met.Arch LCC/PC/Gen/1/25.

21 L.Met.Arch LCC/PC/Gen/1/25.

22 For an interesting analysis of the links between the promotion of electricity and social power in Britain see Luckin (1990).

23 L.Met.Arch LCC/PC/Gen/1/25; L.Met.Arch LCC/PC/Gen/1/33; see also National Smoke Abatement Society (1936).

24 G.Cit.Arch MP40/208; L.Met.Arch LCC/PC/Gen/1/33.

25 L.Met.Arch LCC/PC/Gen/1/14.

26 Note that a significant amount of time was spent during the 1843 Select Committee on Smoke Prevention interviewing engineers, boiler designs and combustion experts on the most effective means of improving coal combustion, HC.PP.1843(583).

27 It is possible to see the standardised scientific training of stokers as undermining their power in the workplace. The scientific discourses of combustion served to work against traditional views of the boiler as an idiosyncratic individual, whose behaviour and preferences could only be effectively interpreted by its attendant operator. Standardised scientific training suggested that any certified person could work efficiently with any piece of combustion technology.

CHAPTER FIVE

1   Committee for the Investigation of Atmospheric Pollution (1916) 'First Report of the Committee of Investigation – Presenting the Results Obtained for the Year April 1914 to March 1915' (Office of the Committee for the Investigation of Atmospheric Pollution, London) [Reprinted from *The Lancet*, 26 February 1916], p. iv. M.Off.Arch.MO 249 256.

2   Thorsheim does, however, note that in the 1880s Glasgow's sanitary department maintained a number of stations that were devoted to the routine assessment of the gaseous content of the air (2006: 127–8).

3   While Braun does not explicitly explore the role of instruments in changing governmental rationality, he does show how the changing insights of structural geologists (which were in part based upon new forms of instrumentation) transformed the Canadian State's system of land law.

4   Committee for the Investigation of Atmospheric Pollution (1916): i, M.Off. Arch.MO 249 256.

5   Ibid.

6   Ibid.

7   To these ends F.J.W. Whipple, the Superintendent of Instruments at the Meteorological Office, and member of the CIAP, played a crucial role in the transfer of technological devices between the established and fledgling sciences.

8   Committee for the Investigation of Atmospheric Pollution (1916): ii, M.Off. Arch.MO 249 256.

9   Within his discussion of the normal and abnormal, Foucault argues that these categories reflect paradigms of behaviour and socioeconomic conditions that respectively enable and inhibit the achievement of pre-established and prescriptive norms. In this context, Foucault claims that it is the predetermined norm, not the normal, that provides the locus for disciplinary government; that is to say that disciplinary government is first about *normation* (the establishment of ideal paradigms of society) and only then *normalisation* (ensuring that modes of social action and organisation facilitate the achievement of norms) (see Foucault (2007 [2004]): 57).

10   Committee for the Investigation of Atmospheric Pollution (1916): ii, M.Off. Arch.MO 249 256.

11   Ibid.: iv.

12   Ibid.: viii.

13   Records of the regular correspondence between the CIAP and local analysts reveal the systems of familiarity which bound this relatively small community of air pollution scientists together, see M.Off.Arch.MO 249 256.

14   See the CIAP 'Discussion of Results', Committee for the Investigation of Atmospheric Pollution (1916): xxviii, M.Off.Arch.MO 249 256.

15   Ibid.

16   For a more detailed account of the problematics of rainfall variation for the monitoring sciences of air pollution see the CIAP 'Discussion of Results', Committee for the Investigation of Atmospheric Pollution (1916): M.Off.Arch. MO 249 256.

17 Ibid.: xxix.
18 Committee for the Investigation of Atmospheric Pollution (1917) 'Second Report of the Committee of Investigation': M.Off.Arch.MO 249 256.
19 Ibid.
20 TNA.DSIR14/1.
21 Committee for the Investigation of Atmospheric Pollution (1917): 3, M.Off. Arch.MO 249 256.
22 Ibid.
23 Ibid.: 14.
24 Ibid.
25 It is important to note that Owens was not the first to devise an effective apparatus for obtaining a sample of suspended air pollution. The chemist Julius B. Cohen (who was an original member of the CIAP) had already devised a device for measuring suspended sulphur pollutants that utilised a solution of hydrogen peroxide, see Thorsheim (2006): 128.
26 Committee for the Investigation of Atmospheric Pollution (1917): 14–15, M.Off.Arch.MO 249 256.
27 Although, see Latour's analysis of the use of the Munsell colour code in the field-based classification of tropical soils (Latour, 1999: 58–60).
28 Committee for the Investigation of Atmospheric Pollution (1917): M.Off.Arch. MO 249 256.
29 Advisory Committee on Atmospheric Pollution (1919) 'Report on Observations in the Year Ending March 31 1919': 22–7, M.Off.Arch.MO 249 256.
30 Ibid.
31 Ibid.: 26.
32 Ibid.
33 Ibid.
34 Advisory Committee on Atmospheric Pollution (1922) 'Report on Observations in the Year Ending March 31 1922: 38, M.Off.Arch.MO 249 256.
35 Ibid.: 38. See also Owens (1922).
36 See Tyndall, J. *Floating Matter of the Air*, M.Off.Arch.MO 249 256.
37 Advisory Committee on Atmospheric Pollution (1922) 'Report on Observations in the Year Ending March 31 1922: 38, M.Off.Arch.MO 249 256.
38 Ibid.
39 Ibid.: 38–9.
40 See Thorsheim (2006) for a discussion of the perceived disinfectant qualities of smoke pollution.

CHAPTER SIX

1 See Elden (2006) for a fascinating discussion of how the spatialised geopolitics of *Lebensraum* informed the political calculations and actions of Nazi territorial expansion and spatial planning in Germany. Elden reveals how and why the types of bureaucratised racial calculations practised by the German Nazi Party were informed and structured by an awareness of spatial measurement and

politics. For a related analysis of the links between eugenic science and carto-graphy see Crampton (2007: 223–44).

2   See record of the Committee of Enquiry into the Future of the Advisory Committee on Atmospheric Pollution, TNA.DSIR14/1.

3   Letter from John Switzer Owens to Air Ministry 17 April 1924, TNA. DSIR14/1.

4   Letter from L.S Lloyd recounting proceedings of meeting chaired by Sir William Nicholson to discuss the future of the ACAP 24 April 1925, TNA. DSIR14/1.

5   TNA.DSIR14/1

6   Ibid.

7   Letter from L.S Lloyd recounting proceedings of meeting chaired by Sir William Nicholson to discuss the future of the ACAP 24 April 1925, TNA. DSIR14/1.

8   TNA.DSIR14/1.

9   Ibid.

10 Letter from L.S. Lloyd to Mr Henry Tizard 4 March 1926, TNA.DSIR14/1. It is interesting to note that Henry Tizard occupied an important point of connec-tion between the DSIR and Air Ministry coordinating as he did the DSIR's Board on Defence Research (see Rose & Rose, 1971: 60).

11 See Moseley (1980) for an interesting discussion of the tensions that existed between the DSIR and the Royal Society over the operations of the National Physical Laboratory in inter-war Britain.

12 TNA.DSIR14/1.

13 Ibid.

14 Letter from L.S. Lloyd to Sir Napier Shaw 15 February 1926, TNA.DSIR14/1.

15 See L.Met.Arch LCC/PC/Gen/1/10.

16 Ibid.

17 Section 10 of the 1926 Public Health (Smoke Abatement) Act made provision for State funding support (up to a limit of £500) for local government research into the effective monitoring of air pollution, see L.Met.Arch LCC/PC/Gen/1/10.

18 Interestingly, the decision that the DSIR should assume responsibility for the ACAP was unofficially made at the Inter-Departmental Conference to discuss the future of the ACAP that was convened at the Air Ministry on 23 April 1925. It appears that the continued delay of the DSIR in assuming control of the ACAP reflects an attempt to cut off all government funding from the Committee – TNA.DSIR14/1. The DSIR also assumed responsibility for administering and funding the grant-in-aid for local authority based research developments in the field of air pollution monitoring pro-visioned within Section 10 of the 1926 Public Health (Smoke Abatement) Act.

19 TNA.DSIR14/2.

20 Ibid.

21 Terms of reference for the Atmospheric Pollution Research Committee, TNA. DSIR14/2.

22 This research was concerned less with monitoring technologies and more with the development of commercially viable pollution abatement devices, TNA. DSIR14/2.

23   TNA.DSIR14/2.
24   Ibid.
25   TNA-HLG/55/32.
26   L.Met.Arch LCC/PC/Gen/1/17.
27   TNA-HLG/55/32.
28   Drawing on the work of Henri Lefebvre, John Pickles (2004) observes that ter-
      ritories do not precede maps, but rather maps creatively envisage, shape and
      produce territories (see Lefebvre, 2000: 84–6). Pickles's notion can, I believe,
      be usefully extended to work on governmentality. In the context of governmen-
      tal cartography, it is clear that maps do not precede governmental rationality,
      but are themselves two-dimensional manifestations of existing governmental
      desires. A question then presents itself: as an artefact of atmospheric govern-
      mentality what can the SCCB map of 1930 tell us about governmental plans for
      air pollution surveillance in inter-war Britain?
29   TNA-HLG/55/32.
30   Ibid.
31   TNA-HLG 55/32.
32   Ibid. In 1949 the DSIR estimated that there were 177 deposit gauges; 9 auto-
      matic filters; 38 smoke filters; 30 volumetric apparatus for measuring dioxides
      of sulphur; 272 lead peroxide instruments; and 12 instruments for measuring
      the intensity of daylight, TNA-HLG 55/32.
33   L.Met.Arch LCC/PC/Gen/1/18.
34   Ibid.
35   Ibid.
36   Letter from Mr John Edwards, London County Council to Chief Officer of
      DSIR 2 October 1939, ibid.
37   Letter from Mr John Edwards, London County Council to Chief Officer of
      DSIR, 1 November 1940, ibid.
38   Ibid.
39   Ibid.
40   Ibid.
41   The continuation of the work of the SCCB throughout the Second World War
      appears to have owed much to the voluntary nature of the Conference.
42   TNA.DSIR14/2.
43   Ibid.
44   M.Off.Arch.09/BF.24.
45   Report in *Abingdon Local Weather Phenomena Book* – M.Off.Arch.09/BF.24.
46   M.Off.Arch.09/BF.24.
47   Ibid. One of the largest scale military studies of this kind was conducted in
      the skies above Boscombe Down, Wiltshire, between January and June 1941.
      Over 69 ascents were conducted from 43,000 feet during this period using
      specially equipped planes. Such studies provided unique insights into the
      temperature gradients and wind speeds at different altitudes in the atmo-
      sphere.
48   For more on the notion of flat ontology see Marston, Jones and Woodward
      (2005). For an interesting analysis of the notion of vertical territoriality that
      focuses on the geo-politics of air power see Williams (2007).

49　In addition to work that has extended Foucault's own analysis of disciplinary spaces like asylums, clinics and prisons, there has been a growing interest in the role of urban planning and design within the government of urban spaces, see the work of Stephen Legg on the formation of disciplinary spaces within the design and building of New Delhi (2007: 82–148); and Huxley's reflections on the formation of model towns (2006: 771–87). See also Foucault's own reflections on the role of urban design in Legg (2007).

50　See Legg (2007: 194)

51　Ibid.

52　British Standards Institution (1951), see L.Met.Arch LCC/PC/Gen/1/18.

53　For a brief, but excellent overview of the London fog disaster, see Thorsheim (2006). See also Greater London Authority (2002) for an interesting retrospective overview of the fog disaster.

54　See Thorsheim (2006: 154). Thorsheim explains that the difficulties associated with accurately determining the deaths associated with the London fog disaster are twofold: (i) the problems of matching the administrative districts used to collect health data with those affected by the fog; and (ii) the problems of statistical comparison: in order to know the true impacts of the fog disaster it is necessary to compare deaths rates with other time periods – some studies compared death rates during the London fog with weeks immediately preceding and following the disaster while others compared death rates over the same dates in other years (2006: 161–3).

55　Whether the fact that the disaster was centred on London, the seat of British government, and not other urban areas in Britain was the reason for such a rapid political response is difficult to assess. Because the disaster did affect London, and both the working and upper classes alike was, however, a clear stimulus for effective government action. The way that the London fog disaster affected both the upper and working classes of London is perhaps seen most clearly in the famous incident of the Sadler's Wells production of *La Traviata* having to be cancelled due to fog entering the theatre and obstructing the audience's vision of the stage, see *The Times* (1952c).

56　The Executive Committee of the National Smoke Abatement Society had called for an 'immediate and intensive' government inquiry into the severe air pollution of December 1952 (*The Times*, 1952d).

57　It is worth noting the change in emphasis between the 1920s and the 1950s concerning what is a legitimate basis for governmental research in the field of air pollution. In the 1920s the DSIR argued that the government should only support research into air pollution that had direct commercial benefits. Following the London fog disaster it became clear that the potential socioeconomic costs of air pollution were so large that greater government support was needed to assist atmospheric pollution studies.

58　Dr Wilkins made this announcement at a special meeting of the Royal Meteorological Society to discuss fog smoke. His assessment was based upon the fact that similar health symptoms were experienced simultaneously by thousands of people over a 100 square mile radius, *The Times* (1953c).

59　For more detail on the history and composition of the Beaver Committee, see Ashby and Anderson (1981): 106–11.

60   Taken from a speech delivered by Sir Hugh Beaver in New York, 2 March 1955, quoted in Thorsheim (2006: 174).
61   There were several notable exceptions to the principles established by the Beaver Committee. These included furnaces that tended to produce dark smoke in the early stages of combustion, and premises that would require significant technological investment before air pollution could be abated (Ashby & Anderson,1981: 107).
62   Note, however, that following the initial publication of the Beaver Committee's Report the government was still not willing to produce new legislation. It was only after Gerald Nabarro won a ballot for a private members bill on pollution abatement that the government stepped in (see Greater London Authority, 2002: 15).
63   The Beaver Committee recommended the use of the *Ringelmann chart* as a way of visually recognising dark smoke. The Ringelmann chart had been used previously to control smoke production in the city of Pittsburgh, see Ashby and Anderson (1981: 107–8).
64   Smokeless zones were areas within which the production of smoke was completed prohibited. Smoke control areas, on the other hand, were areas within which only smoke produced from government-approved fuels was permitted (Greater London Authority, 2002: 15).
65   The British standard measurement for air pollution was part of the same British Standard (BS1747) as that previously discussed for deposit gauges. This guidance was issued as Part II of the BS1747.
66   For a detailed discussion of sampling methods and errors in sampling and measurement, see Warren Spring Laboratory (1972b): 111–18.
67   For the use of comparative regional data sets see Warren Spring Laboratory (1972a): 15–18.

CHAPTER SEVEN

1   See MIT Libraries, http://libraries.mit.edu/archives/exhibits/energy/index.html (accessed 4 February 2008). The final report of the Project on the Predicament of Mankind was published in 1972 as Meadows *et al.* (1972).
2   Minute of meeting convened by the Department of the Environment to discuss aspects of pollution, 13 July 1971, TNA–LG1/532/16.
3   From the home page of the Royal Commission on Environmental Pollution, http://www.rcep.org.uk/ (accessed 29 May 2008).
4   The ICAPR was established following the Beaver Report in 1954. The Committee was designed to better harmonise air pollution research in different government departments and to support the existing work of the Atmospheric Pollution Research Committee of the Fuel Research Station.
5   Notes of a meeting held by the Interdepartmental Committee on Air Pollution Research, 22 November 1971, TNA–LG1/532/16.
6   TNA–LG1/532/16.
7   Notes of a meeting held by the Interdepartmental Committee on Air Pollution Research, 21 June 1971, TNA–LG1/532/16.

8   Working Party on Air Pollution Monitoring – Draft Report, TNA–LG1/532/16.
9   TNA–LG1/532/16.
10 Letter from M.W. Holdgate, dated 28 October 1971, TNA–LG1/532/16.
11 Letter from R.G. Adams to Graham Fuller, dated 8 September 1971, TNA–LG1/532/16.
12 TNA–LG1/532/16.
13 Advisory Committee on Atmospheric Pollution (1918). *Report of Observations in the Year 1917–1918* (pp. 20–3). London: Office of the Advisory Committee on Atmospheric Pollution. M.Off.Arch.MO 249 256.
14 UK Air Quality Archive (2008), http://www.airquality.co.uk/archive/monitoring_networks.php?n=history (accessed 19 February 2008).
15 Ibid.
16 This organisation was formerly called the National Environment Technology Centre (NETCEN).
17 For more on the growth and expansion of computer-based modelling sciences and their application to different aspects of atmospheric enquiry see Norton and Suppe (2001).
18 Note, however, that with the emergence of new statutory requirements for key polluters to directly monitor their air emissions, the NETCEN is attempting to move gradually away from pollution estimates to directly monitored data.
19 Personal interview conducted with representative of Defra's statistical division, May 2004.
20 Personal interview conducted with representative of Defra's statistical division, May 2004.
21 *Infostructures* is a term deployed by Luke to describe the cyberspatial structures which are increasingly replacing the infrastructures of state bureaucracies, ibid.
22 See Defra (2004: 7) for an overview of health-related air quality objectives in operation in the UK.
23 Supported by the United Nations Economic Commission for Europe, the Aarhus Convention seeks to secure the rights of citizens to gain information concerning environmental change and pollution as a way of securing greater participation in the sustainable development process and associated forms of policy development.
24 The UK Air Quality Archive is available at http://www.airquality.co.uk/archive/index.php.
25 For more information about this particular sampling site go to http://www.bv-aurnsiteinfo.co.uk/viewSite.asp?pageRef=151&stationID=81 (accessed 4 March 2008).

CHAPTER EIGHT

1   Ironically, it appears that the distinctive and lethal radioactive traces left from nuclear experiments made it easy for scientists to study the different metabolic interactions of the very ecosystems that were being threatened by nuclear devices.

2   See, for example, Foucault's discussion of the repressive sexual hypothesis and the importance of placing it within a general history of sexual discourse rather than utilising as a causal mechanism within accounts of sexual conduct and politics (Foucault, 1998 [1976]: 3–13).

3   From the home page of the Royal Commission on Environmental Pollution, http://www.rcep.org.uk/ (accessed 29 May 2008).

4   Ibid.

5   For more on the role of eco-modernist thinking in British environmental pollution control see Weale (1992).

6   The UNECE is one of five regional commissions for the UN. The other four are: the Economic and Social Commission for Asia and the Pacific; the Economic Commission for Latin America and the Caribbean; the Economic Commission for Africa; and the Economic and Social Commission for Western Asia.

7   Although the notion of the biosphere has long historical antecedents, stretching into the nineteenth century, it was the Russian scientist V.I. Vernadskii who is most associated with the systematic development of the idea. For more details on Vernadskii and the biosphere concept see Oldfield and Shaw (2006).

8   The ADMN is made up of two networks. The primary network aims to provide high quality information on the changing temporal quantities of acid deposited in Britain. The secondary network is dedicated to providing a record of the spatial deposition of acid deposition, see DETR (2001).

9   As with the Automated Urban and Rural Network (see previous chapter), the ADMN is maintained and administered by the AEA Energy and Environment on behalf of Defra.

10 Of the 210 sampling sites 35 did not contain any of the four moss species (Ashby et al., 2002: 2).

11 For more on the precise protocols of moss storage and analysis see Ashby et al., 2002: 11–13.

12 The Gothenburg Protocol is also referred to as the multi-effect protocol and was ratified in May 2005.

13 In a related sense, Foucault described how the government of grain scarcity and fluctuation provided a key context for biopolitical calculation, and from the eighteenth century onwards an early arena for governmentality, see Foucault, 1994 [2007]: 30–49.

14 According to Rutherford (1999: 55–60) the rise of ecological rationality within the mainstream practices of government is most evident in the emergence of environmental impact assessment procedures and associated planning practices.

CHAPTER NINE

1   See Nick Cohen's recent reflections on Foucault's relationship with the Ayatollah Khomeini and his views on the Iranian revolution (Cohen, 2007: 107–8).

2 For more on the place of the 1978 lecture series within Foucault's broader political life see Michel Sennellart's 'Course Context' in Foucault (2007 [2004]): 369–401.

3 See also Deleuze's (1999 [1986]: 21–38) thought-provoking characterisation of Foucault as a *cartographer*.

4 '*Omnes et Singulatim*: Towards a Critique of Political Reason' was the of Foucault's two Tanner lectures. These lectures were given at Stanford University on 10 and 16 October 1979. While apologising for its 'pretentious' title, these lectures drew together Foucault's reflections on the rationalities of governmental power he had been developing through his lecture courses at the Collège de France. Particular attention was given to the connections between the simultaneous government of a population and each subject of that population. While this problematic was itself defined in the notion of pastoral power (and the role of the shepherd) these lectures appear to represent an important bridge between Foucault's concern with aggregate forms of governmental power and his later analysis of the care of the self.

5 Agrawal develops the concept of *intimate government* in direct contrast to notions of government at a distance. According to Agrawal, intimate government differs from *government at a distance* in that while governing at a distance involves the maintenance of strong connections between a control centre of knowledge calibration and coordination and a locality, intimate forms of government see governmental practices become more of a decentred set of norms within the communities being governed (2005: 195).

6 See Monbiot (2006) for a fascinating analysis of the emergence of the climate change denial industry and a particular interesting, if worrying, analysis of its connections with the pro-smoking lobby.

7 While carbon dioxide is not recorded systematically through direct monitoring it is worth noting that other greenhouse gases (covered in the Marrakesh Accords) are recorded as part of Britain's network of air monitoring devices. These indirect greenhouse gases include nitrogen oxide, carbon monoxide and sulphur dioxide.

8 For more on the statistical processes that inform the construction of the *UK's Greenhouse Gas Inventory* go to http://www.ghgi.org.uk (accessed 31 July 2008).

9 For more information on estimates of British carbon emissions go to the Carbon Trust website, http://www.carbontrust.co.uk/default.ct (accessed 14 August 2008).

10 http://www.defra.gov.uk/environment/climatechange/uk/index.htm (accessed 10 December 2008).

11 While the latest IPCC *Synthesis Report* (2007) suggests that the chance of *large-scale singularities* (or abrupt changes at a high order of magnitude) are unlikely, it does recognise that dynamic processes of environmental feedback could contribute to faster than anticipated changes being brought on by climate change (see, for example IPCC, 2007: 65 for a discussion of large-scale singularities and the associated results between sea-level changes and Arctic ice sheets).

12 See in particular the Community Carbon Footprinting initiative developed by Herefordshire County Council, http//:www.myherefordshire.com/carbonfootprinting.aspz (accessed 1 August 2008).

13   This initiative is supported by the US Environmental Protection Agency's *EMPACT: Environmental Monitoring for Public Access and Community Tracking* programme. For more information on other atmospheric projects operating within this scheme go to: http://www.epa.gov/empact/air.htm (accessed 1 August 2008).

14   See Latour (2004: 207) for an interesting discussion of the value and importance of forging new *learning compacts* to explore the complex interconnections between humans and the environment. According to Latour, learning compacts could usefully replace social contracts in asserting '[t]he common ignorance of the governors and the governed in a situation of collective experimentation' (ibid.: 243).

References

ARCHIVAL SOURCES

Birmingham City Archives: Birmingham City Council

(Held at Birmingham Central Library)
B.Cit.Arch BCC/AR-74 Borough Inspection Committee (1856–1873) – Minute Book 74.
B.Cit.Arch BCC/AR-76 Borough Inspection Committee (1856–1873) – Minute Book 76.
B.Cit.Arch BCC/AR-78 Borough Inspection Committee (1856–1873) – Minute Book 78.
B.Cit.Arch BCC/AR-72 Sanitary Committee (1872–1911) – Minute Book 72.
B.Cit.Arch BCC/AR-73 Sanitary Committee (1872–1911) – Minute Book 7.

Glasgow City Archives

(Held at the Mitchell Library, Glasgow)
G.Cit.Arch MP40/208 Newspaper cuttings from Clean Air Exhibition of 1910.
G.Cit.Arch P.175 Photograph 175 of Glasgow City Archive Collection.
G.Cit.Arch P.177 Photograph 177 of Glasgow City Archive Collection.
G.Cit.Arch P.183 Photograph 183 of Glasgow City Archive Collection.
G.Cit.Arch DTC.14.2 (6) Local government reports and pamphlets (1862–1899).

House of Commons: Parliamentary Papers

HC.PP.1843(583)-Final Report Select Committee on Smoke Prevention, 1843, *Final Report*.
HC.PP.1843(583)-App.6 Select Committee on Smoke Prevention, 1843, *Police Report showing the results of four hours observation commencing at two o'clock and ending at four pm each day, on the smoke issuing from 53 factory chimneys casually selected from 462 chimneys*, Appendix 6.

HC.PP.1846 (194) Smoke Prohibition Report, Addressed to Viscount Canning, Sir Henry Thomas De la Bech and Dr Lyon Playfair, *The Means of obviating the Evils arising from the SMOKE occasioned by Factories and other Works situated in large Towns*.

London Metropolitan Archives: London County Council

(Farringdon, London)

L.Met.Arch LCC/PC/Gen/1/9 General papers (1925–1926).

L.Met.Arch LCC/PC/Gen/1/10 Details of bylaws passed and other local government responses to the 1926 Public Health (Smoke Abatement) Act.

L.Met.Arch LCC/PC/Gen/1/14 Atmospheric Pollution and Smoke Abatement (reports of smoke from railway locomotives and general smoke observation reports) (1923–1947).

L.Met.Arch LCC/PC/Gen/1/17 Notes and general correspondence detailing liaison between the London County Councils and Standing Conference on Cooperating Bodies (1935–1947).

L.Met.Arch LCC/PC/Gen/1/18 Correspondence concerning the supply, use and replacement of air pollution monitoring equipment between the London County Councils and Standing Conference on Cooperating Bodies (1919–1945).

L.Met.Arch LCC/PC/Gen/1/20 Deposit gauge records, London County Council.

L.Met.Arch LCC/PC/Gen/1/25 Atmospheric Pollution and Smoke Abatement (General Files) containing press cuttings and journal entries reviewing various smoke abatement exhibitions (1912–1956).

L.Met.Arch LCC/PC/Gen/1/33 Atmospheric Pollution and Smoke Abatement (General Files) containing information about Alkali Acts, smoke inspection, smokeless home exhibitions (1878–1914).

Meteorological Office Archives

(Held at the Devon Records Office, Exeter)

M.Off.Arch MO 249 256 Committee for the Investigation of Atmospheric Pollution (Renamed the Advisory Committee on Atmospheric Pollution).

M.Off.Arch 09/BF.24 Synoptic Division Technical Memoranda (Air Ministry).

The National Archives

(Kew, London)

TNA.HLG/55/209 National Smoke Abatement (General Files) containing information regarding application of society for government funding, minutes of annual meetings and National Air Pollution Questionnaire.

TNA.DSIR14/1 Committee minutes, memoranda and letters associated with the formation, role and decisions of the Committee of Enquiry into the Future of the Advisory Committee on Atmospheric Pollution.

TNA.DSIR14/2 Correspondence and terms of reference associated with the formation of the DSIR's Atmospheric Pollution Research Committee and the Standing Conference of Cooperating Bodies.

TNA.HLG/55/32 Committee minutes, notes and correspondence of the Standing Conference of Cooperating Bodies.

TNA.LG1/532/16 Minutes and correspondence of the Interdepartmental Committee on Air Pollution Research – The Future of the National Survey.

UNPUBLISHED SOURCES

Birmingham Borough Inspection Committee (1872). Report on the Public Health Act 1872, the Steam Whistles Act and the Adulteration of Food etc., Act for Presentation to the Council. Steam Printing Offices, Birmingham.

Birmingham Sanitary Committee (1875). Report of the Sanitary Committee for Presentation at a Special Meeting of the Council. 16 February 1875.

PUBLISHED PAMPHLETS, CIRCULARS, GUIDES AND HANDBOOKS

British Standards Institution (1951). *Deposit Gauges for Atmospheric Pollution*. London: British Standards Institution.

Clinch, H.G. (1923). *The Smoke Inspector's Handbook – or Economic Smoke Abatement*. London: H.K. Lewis & Co. Ltd.

Dickinson, E. (1929). *Successful Stoking and Smoke Abatement: A Manual for Boiler Attendants*. Wakefield, Sanderson & Clayton Printers.

The National Smoke Abatement Society (1936). *Smoke Abatement Exhibition Handbook and Guide*. Manchester: The National Smoke Abatement Society.

PUBLISHED BOOKS, ARTICLES AND REPORTS

Agamben, G. (1998). *Homo Sacer: Sovereign Power and Bare Life*. Stanford, CA: Stanford University Press.

Agrawal, A. (2005). *Environmentality: Technologies of Government and the Making of Subjects*. Durham, NC: Duke University Press.

Air Quality Archive (2008). A history of air pollution monitoring in the UK, http://www.airquality.co.uk/archive/monitoring_networks.php?n=history (accessed 19 February 2008).

Anderson, K. (2003). White natures: Sydney's Royal Agricultural Show in post-human perspective. *Transactions of the Institute of British Geographers*, **28**: 422–41.

Ashby, E. & Anderson, M. (1981). *The Politics of Clean Air*. Oxford: Clarendon Press.

Ashmore, M., Bell, S., Fowler, D. *et al.* (2002) Survey of UK metal content in mosses 2000. Part 2 of Final Contract Report EPG.

Barry, A. (1998). Motor ecology: the political chemistry of urban air. Centre for Urban and Community Research Occasional Paper Series. Goldsmiths, University of London.

Barry, A. (2001). *Political Machines: Governing a Technological Society*. London: Athlone.

Barthes, R. (2000). *Camera Lucida*. London: Vintage.

Baudrillard, J. (1981[1994]). *Simulation and Simulacra* (Trans. S.F. Glaser). Michigan: University of Michigan Press.

Bennett, T. (1995). *The Birth of the Museum: History, Theory, Politics*. London: Routledge.

Berger, J. (1972). *Ways of Seeing*. London: Penguin.

Berman, M. (1983). *All that is Solid Melts into Air: The Experience of Modernity*. London: Verso.

Blissett, M. (1972). *Politics in Science*. Boston: Little, Brown.

Botkin, D.B. (1992). *Discordant Harmonies: A New Ecology for the Twenty-First Century*. Oxford: Oxford University Press.

Braun, B. (2000) Producing vertical territory: geology and governmentality in late Victorian Canada. *Ecumene*, 7: 7–46.

Brimblecombe, P. (1987). *The Big Smoke: A History of Air Pollution in London since Medieval Times*. London: Methuen.

Brimblecombe, P. (2004). Perceptions and effects of late Victorian air pollution. In E.M. DuPuis (ed.), *Smoke and Mirrors: The Politics and Culture of Air Pollution* (pp. 15–26). London: New York University Press.

Bristow, M. (2008). Lingering pollution worries China. British Broadcasting Corporation. 29 July, http://news.bbc.co.uk/1/hi/world/asia-pacific/7528523.stm (accessed 29 July 2008).

British Broadcasting Corporation (2008). Beijing pollution: facts and figures. 16 July, http://news.bbc.co.uk/1/hi/world/asia-pacific/7498198.stm (accessed 11 August 2008).

Brown, W. (1994). Deaths linked to London smog. *New Scientist*, 1931 (5 June), 4.

Burchell, G., Gordon, C. & Miller, P. (eds) (1991). *The Foucault Effect: Studies in Governmentality*. London: Harvester Wheatsheaf.

Butler, J. (1990). *Gender Trouble: Feminism and the Subversion of Identity*. London: Routledge.

Callon, M. & Latour, B. (1981). Unscrewing the big Leviathan: how actors macro-structure reality and how sociologists help them to do so. In K.D. Knorr-Centina & A.V. Cicourel (eds), *Towards an Integration of Micro- and Macro-Sociologies* (pp. 277–303). London: Routledge & Kegan Paul.

Carson, R. (2000). *Silent Spring*. London: Penguin.

Centre for Ecology and Hydrology (2001). *The Ozone Umbrella Project – Final Report*. Huntingdon: Centre for Ecology and Hydrology.

Centre for Ecology and Hydrology (2004). *Update: The Status of UK Critical Loads, Critical Loads Methods, Data and Maps*. Huntingdon: Centre for Ecology and Hydrology.

Centre for Ecology and Hydrology (2008) What is a critical load? http://critloads.ceh.ac.uk/what_is_cl.htm (accessed 31 March 2008).

Charbit, Y. (2002). The political failure of an economic theory: physiocracy. *Population*, 57, 855–83.

Cohen, N. (2007). *What's Left – How Liberals Lost Their Way*. London: Fourth Estate.

Crampton, J.W. (2007). Maps, race and Foucault: eugenics and territorialization following World War I. In J.W. Crampton & S. Elden (eds), *Space, Knowledge, Power: Foucault and Geography* (pp. 223–44). Aldershot: Ashgate.

Crampton, J.W. & Elden, S. (2006). Space, politics, calculation: an introduction. *Social and Cultural Geography*, 7: 681–5.

Crampton, J.W. & Elden, S. (2007) (eds). *Space, Knowledge, Power: Foucault and Geography*. Aldershot: Ashgate.

Crary, J. (1992). *Techniques of the Observer: On Vision and Modernity in the Nineteenth Century*. Cambridge, MA: MIT Press.

Crary, J. (2000). *Suspensions of Perception: Attention, Spectacle and Modern Culture*. Cambridge, MA: MIT Press.

Cresswell, T. (2006). *On the Move: Mobility in the Modern West*. New York: Routledge.

Darier, É. (1990). Foucault and the environment: an introduction. In É. Darier (ed.), *Discourses of the Environment* (pp. 1–33). Oxford: Blackwell.

DECDEP (1984). *Controlling Pollution: Principles and Prospects (the Government's Response to the Tenth Report of the Royal Commission on Environmental Pollution)*. Pollution Paper 22. London: HMSO.

DETR (2001). *Operation and Management of the UK Acid Deposition and Monitoring Networks*, AEA/ENV/R/0523 (Issue 2).

Defra (2004). UK air pollution (brochure). Oxford: NETCEN.

Defra (2006). *The Air Quality Strategy for England, Scotland, Wales and Northern Ireland: A Consultation Document on Options for Further Improvements in Air Quality*. London: Defra.

Defra (2008). UK air quality forecasting: A UK particulate episode from 24th March to 2nd April 2007. AEAT/Env/R/2566, 1 (January).

Dean, M. (1999). *Governmentality: Power and Rule in Modern Society*. London: Sage.

Debord, G. (1992). *Society of Spectacle*. Rebel Press.

Deleuze, G. (1999 [1986]). *Foucault* (Trans. S. Hand). London: Continuum.

Deleuze, G & Guittari, F. (1987) *A Thousand Plateaus: Capitalism and Schizophrenia* (Trans. B. Massumi). Minneapolis: University of Minnesota Press.

Department of the Environment (1974). *Monitoring the Environment in the UK – A Report by the Central Unit on Environmental Pollution*. London: HMSO.

Department of the Environment (CUEP) (1975). *Controlling Pollution: A Review of Government Action Related to Recommendations by the Royal Commission on Environmental Pollution*. London: HMSO.

Dillon, M. (2004). Correlating sovereignty and biopower. In J. Edkins & V. Pin-Fat (eds), *Sovereign Lives: Power in Global Politics* (pp. 41–60). London: Routledge.

Dixon, D. & Jones, J.P. III. (1996). For a supercalifragilisticexpialidocious scientific geography. *Annals of the Association of American Geographers*, 86: 767–79.

Douglas, M. (1966). *Purity and Danger: An Analysis of Concepts of Pollution and Taboo*. London: Routledge.

Driver, F. (1988). Moral geographies: social science and the urban environment in mid-nineteenth-century England. *Transactions of the Institute of British Geographers*, 13: 275–87.

DSIR (1928). *A Note on the Investigation of Atmospheric Pollution*. London: HMSO.

DuPuis, E.M. (2004). Introduction. In E.M. DuPuis, (ed.), *Smoke and Mirrors: The Politics and Culture of Air Pollution* (pp. 1–11). New York: New York University Press.

The Economist (2008). The electronic bureaucrat: A special report on technology and government. 16 February.

Edney, M. (1999). *Mapping Empire: Geographical Construction of British India, 1763–1843*. Chicago: University of Chicago Press.

Elden, S. (2001). *Mapping the Present: Heidegger, Foucault and the Project of a Spatial History*. London: Continuum Press.

Elden, S. (2006). National socialism and the politics of calculation. *Social and Cultural Geography*, 7: 753–69.

Elden, S. (2007). Governmentality, calculation, territory. *Environment and Planning D: Society and Space*, 25: 562–80.

Endfield, G.H. (2007). Archival explorations of climatic variability and social vulnerability in colonial Mexico. *Climatic Change*, 83: 9–38.

Endfield, G.H. (2008). *Climate and Society in Colonial Mexico: A Study in Vulnerability*. Oxford: Blackwell.

Environmental Protection Agency (2002). Planning and implementing a real-time air pollution monitoring and outreach programme for your community: the AirBeat Project of Roxbury, Massachusetts. United States Environmental Protection Agency, Washington, DC (EPA/625/R-02/012).

Foucault, M. (1990 [1984]). *The Care of the Self: The History of Sexuality, Volume 3* (Trans. R. Hurley). London: Penguin.

Foucault, M. (1991 [1975]). *Discipline and Punish: The Birth of the Prison* (Trans. A. Sheridan). London: Penguin.

Foucault, M. (1991). Questions of method. In G. Burchell, C. Gordon & P. Miller (eds), *The Foucault Effect: Studies in Governmentality* (pp. 73–86). London: Harvester Wheatsheaf.

Foucault, M. (1992 [1984]). *The Use of Pleasure: The History of Sexuality, Volume 2*. (Trans. R. Hurley). London: Penguin.

Foucault, M. (1998 [1976]). *The Will to Knowledge: The History of Sexuality, Volume 1* (Trans. R. Hurley). London: Penguin.

Foucault, M. (2000a [1994]). The birth of social medicine in reason. In J.D. Faubion (ed.), *Michel Foucault – Power: Essential Works of Foucault 1954–1984* (pp. 134–56). London: Penguin.

Foucault, M. (2000b [1979]). The political technologies of individuals. In J.D. Faubion (ed.), *Michel Foucault – Power: Essential Works of Foucault 1954–1984* (pp. 403–17). London: Penguin.

Foucault, M. (2000c [1979]). 'Omnes et singulatim': toward a critique of political reason. In J.D. Faubion (ed.), *Michel Foucault – Power: Essential Works of Foucault 1954–1984* (pp. 298–325). London: Penguin.

Foucault, M. (2002 [1961]). *Madness and Civilization: A History of Insanity in the Age of Reason* (Trans. R. Howard). London: Routledge.

Foucault, M. (2003a [1963]). *The Birth of the Clinic. An Archaeology of Medical Perception* (Trans. A.M. Sheridan). London: Routledge.

Foucault, M. (2003b [1966]). *The Order of Things: An Archaeology of the Human Sciences*. London: Routledge.

Foucault, M. (2004 [1997]). Society Must be Defended – Lectures at the Collège de France 1975–1976 (Trans. D. Macey). London: Penguin.

Foucault, M. (2006 [1969]). *The Archaeology of Knowledge* (Trans. A.M. Sheridan) London: Routledge.

Foucault, M. (2007 [2004]). *Security, Territory, Population – Lectures at the Collège de France 1977–1978* (Ed. M. Senellart. Trans. G. Burchell). Basingstoke: Palgrave Macmillan.

Foucault, M. (2008 [2004]). *The Birth of Biopolitics: Lectures at the Collège de France 1978–1979* (Ed. M. Senellart. Trans. G. Burchell). Basingstoke: Palgrave Macmillan.

Gandy, M. (1999). Rethinking the ecological leviathan: environmental regulation in an age of risk. *Global Environmental Change*, **9**: 59–69.

Golinski, J. (2006). *British Weather and the Climate of Enlightenment*. Chicago: University of Chicago Press.

Greater London Authority (2002). *50 Years On: The Struggle for Air Quality in London since the Great Smog of December 1952*. London: Greater London Authority.

Hacking, I. (1990). *The Taming of Chance*. Cambridge: Cambridge University Press.

Hacking, I. (2005 [1983]). *Representing and Intervening: Introductory Topics in the Philosophy of Natural Science*. Cambridge: Cambridge University Press.

Hall, J., Broughton, R., Heywood, E. *et al*. (2001). *Final Report: National Critical Loads Mapping Programme Phase Three*. Huntingdon: Centre for Ecology and Hydrology.

Hannah, M. (2000). *Governmentality and the Mastery of Territory in Nineteenth-Century America*. Cambridge: Cambridge University Press.

Hannah, M. (2008). Biophilia and futurism: oppositional thinking in a biopolitical age. Mimeograph.

Haraway, D. (1991). *Simians, Cyborgs and Women: The Reinvention of Nature*. New York: Routledge.

Huxley, M. (2006). Spatial rationalities: order, environment, evolution and government. *Social and Cultural Geographies*, **7**: 771–87.

Huxley, M. (2007). Geographies of governmentality. In J.W. Crampton & S. Elden (eds), *Space, Knowledge and Power: Foucault and Geography* (pp. 185–204). Aldershot: Ashgate.

IPCC (2007). *Climate Change 2007: Synthesis Report*. IPCC.

Jankovic, V. (2000). *Reading the Skies: A Cultural History of English Weather, 1650–1820*. Chicago: Chicago University Press.

Jenner, M. (1995). The politics of London air: John Evelyn's *Fumifugium* and the Restoration. *Historical Journal*, **38**: 535–51.

Jones, R. (2007). *People/States/Territories: The Political Geographies of British State Transformation*. Oxford: Blackwell.

Jordan, A. (2000). The Europeanization of UK environmental policy, 1970–2000. Working Paper 11/00.

King, K., Sturman, J. & Passant, N. (2006). NAEI UK emissions mapping methodology 2003: A report of the National Atmospheric Emission Inventory (NETCEN).

Klingle, M. (2007). *Emerald City: An Environmental History of Seattle*. New Haven: Yale University Press.

Larner, W. & Le Heron, R. (2005). Neoliberalising spaces and subjectivities: reinventing New Zealand universities. *Organization*, **12**: 843–62.

Latour, B. (1988 [1984]). *The Pasteurization of France* (Trans. A. Sheridan & J. Law). Cambridge, MA: Harvard University Press.

Latour, B. (1993). *We Have Never Been Modern* (Trans. C. Porter). Harlow: Harvester Wheatsheaf.

Latour, B. (1999). *Pandora's Hope: Essays on the Reality of Science of Studies*. Cambridge, MA: Harvard University Press.

Latour, B. (2004). *Politics of Nature: How to Bring the Sciences into Democracy* (Trans. C. Porter). Cambridge, MA: Harvard University Press.

Latour, B. (2006). *Reassembling the Social: An Introduction to Actor–Network Theory.* Oxford: Oxford University Press.

Latour, B. (2007a). A plea for earthly sciences. Key note lecture for the Annual Meeting of the British Sociological Association, London.

Latour, B. (2007b). How to think like a State. Lecture delivered on 22 November 2007 to celebrate the Anniversary of the WWR.

Latour, B. & Woolgar, S. (1986 [1979]). *Laboratory Life: The Construction of Scientific Facts.* Princeton: Princeton University Press.

Lefebvre, H. (2000). *The Production of Space* (Trans. D. Nicholson-Smith). Oxford: Blackwell.

Legg, S. (2007). *Spaces of Colonialism: Delhi's Urban Governmentalities.* Oxford: Blackwell.

Livingstone, D. (2005). Talk, text and testimony: geographical reflections on scientific habits. An afterword. *British Journal for the History of Science,* **38**: 93–100.

Luckin, B. (1990). *Questions of Power: Electricity and Power in Inter-war Britain.* Manchester: Manchester University Press.

Llewellyn, M. (2004). Designed by women and designing women: gender, planning and the geographies of the kitchen in Britain 1917–1946. *Cultural Geographies,* **10**: 42–60.

Luke, T.W. (1995). Simulated sovereignty, telemetric territoriality: the political economy of cyberspace. Presented to the Second Theory, Culture and Society Conference, Culture and Identity: City, Nation, World. 10–14 August.

Luke, T.W. (1999). Environmentality as green governmentality. In É. Darier (ed.), *Discourses of the Environment* (pp. 121–51). Oxford: Blackwell.

McNeill, J. (2000). *Something New under the Sun: An Environmental History of the Twentieth Century.* London: Penguin.

Marston, S.A., Jones, J.P. III & Woodward, K. (2005). Human geography without scale. *Transactions of the Institute of British Geographers,* **30**: 416–32.

Massey, D. (1999). Space-time, 'science' and the relationship between physical and human geography. *Transactions of the Institute of British Geographers,* **24**: 261–76.

McCormick, J. (1991). *British Politics and the Environment.* London: Earthscan.

Meadows, D.H., Randers, J., Meadows, D.L. & Behrens, W.W. (1972). *The Limits to Growth: A Report for the Club of Rome's Project on the Predicament of Mankind.* New York: Universe Books.

Merchant, C. (1983). *The Death of Nature: Women, Ecology and the Scientific Revolution.* San Francisco: Harper.

Mieck, I. (1990). Reflections on a typology of historical pollution: complementary conceptions. In P. Brimblecombe & C. Pfister (eds), *The Silent Countdown* (pp. 73–80). Berlin: Springer Verlag.

Miller, C.A. (2001). Hybrid management: boundary organizations, science policy, and environmental governance in the climate regime. *Science, Technology & Human Values,* **26**: 478–500.

Miller, C.A. & Edwards, P.N. (2001a). Introduction: the globalization of climate science and climate politics. In C.A. Miller & P.N. Edwards (eds), *Changing the Atmosphere: Expert Knowledge and Environmental Governance* (pp. 1–30). Cambridge, MA: MIT Press.

Miller, C.A. & Edwards, P.N. (eds) (2001b). *Changing the Atmosphere: Expert Knowledge and Environmental Governance*. Cambridge, MA: MIT Press.

Miller, P. & Rose, N. (1990). Governing economic life. *Economy and Society*, **19**: 1–31.

Mitchell, T. (2006). Society, economy, and the state effect. In A. Sharma & A. Gupta (eds), *The Anthropology of the State: A Reader* (pp. 169–86). Oxford: Blackwell.

Monbiot, G. (2006). *Heat: How to Stop the Planet Burning*. London: Allen Lane.

Morus, I.R. (1998). *Frankenstein's Children: Electricity, Exhibition, and Experiment in Early-Nineteenth Century London*. Princeton, NJ: Princeton University Press.

Moseley, R. (1980). *Government Science and the Royal Society: The Control of the National Physical Laboratory in the Inter-War Years* (pp. 167–93). Notes and Records of the Royal Society of London.

Mosley, S. (2001). *The Chimney of the World: A History of Smoke Pollution in Victorian and Edwardian Manchester*. Cambridge: The White Horse Press.

Naylor, S. (2002). The field, the museum and the lecture hall: the spaces of natural history in Victorian Cornwall. *Transactions of the Institute of British Geographers*, **27**: 494–513.

Naylor, S. (2006). Nationalizing provincial weather: meteorology in nineteenth-century Cornwall. *British Journal for the History of Science*, **39**: 1–27.

Norton, S.D. & Suppe, F. (2001). Why atmospheric modeling is good science. In C.A. Miller & P.N. Edwards (eds), *Changing the Atmosphere: Expert Knowledge and Environmental Governance* (pp. 67–105). Cambridge, MA: MIT Press.

Oldfield, J.D. & Shaw, D.J.B. (2006). V.I. Vernadskii and the noosphere concept: Russian understandings of society nature interactions. *Geoforum*, **37**: 145–54.

Ong, A. (2005). Ecologies of expertise: assembling flows, managing citizenship. In A. Ong & S.J. Collier (eds), *Global Assemblages: Technology, Politics and Ethics as Anthropological Problems* (pp. 337–53). Malden, MA: Blackwell.

Ong, A. (2007). Neoliberalism as a mobile technology. *Transactions of the Institute of British Geographers*, **32**: 3–8.

O'Riordan, T. (1999). Critical loads. In D.E. Alexander & R.W. Fairbridge (eds), *Encyclopedia of Environmental Science* (p. 102). Klewer.

Orwell, G. (1937). *The Road to Wigan Pier*. London: Penguin.

Owens, J.S.A. (1922). Suspended impurity in the air. *Proceedings of the Royal Society*, **101**.

Owens, S. & Rayner, T. (1999). 'When knowledge matters': the role and influence of the Royal Commission on Environmental Pollution. *Journal of Environmental Policy and Planning*, **1**: 7–24.

Pearce, F. (2007). *The Last Generation: How Nature Will Take Her Revenge for Climate Change*. London: Eden Project Books.

Philo, C. (1992). Foucault's geographies. *Environment and Planning D: Society and Space*, **10**: 137–61.

Pickles, J. (2004). *A History of Spaces: Cartographic Reason, Mapping and the Geo-Coded World*. London: Routledge.

Popper, K. (2002 [1950]). *The Logic of Scientific Discovery*. London: Routledge.

Rabinow, P. & Rose, N. (2006). Biopower today. *BioSocieties*, **1**, 195–217.

Randalls, S. & Thornes, J.E. (2007). Commodifying the atmosphere: 'pennies from heaven'. *Geografiska Annaler Series A: Physical Geography*, **89**: 273–85.

Rhüling, X. & Tyler, G. (1968). An ecological approach to the lead problem. *Botaniska Notiser*, **121**: 321–42.

Rose, G. (1992). Geography as a science of observation: the landscape, the gaze and masculinity. In F. Driver & G. Rose (eds), *Nature and Science: Essays on the History of Geographical Knowledge*. Historical Geography Research Series, No. 28.

Rose, H. & Rose, S. (1971). *Science and Society*. London: Penguin.

Rose, N. (1999a). *Powers of Freedom: Reframing Political Thought*. Cambridge: Cambridge University Press.

Rose, N. (1999b). *Governing the Soul: The Shaping of the Private Self*. London: Free Association Books.

Rose, N. (2007). *The Politics of Life Itself: Biomedicine, Power and Subjectivity in the Twenty-First Century*. Woodstock: Princeton University Press.

Rose, N. & Miller P. (1992). Political power beyond the state: problematics of government. *British Journal of Sociology*, **43**: 173–205.

Royal Commission on Environmental Pollution (1976). *Fifth Report: Air Pollution Control: An Integrated Approach*. London: HMSO.

Royal Commission on Environmental Pollution (2007). *Twenty-Sixth Report: The Urban Environment*. London: HMSO.

Rutherford, P. (1999). The entry of life into history. In É. Darier (ed.), *Discourses of the Environment* (pp. 37–62). Oxford: Blackwell.

Rutherford, S. (2007). Green governmentality: insights and opportunities in the study of nature's rule. *Progress in Human Geography*, **31**: 291–307.

Schaffer, S. (1992). A manufactory of OHMS, Victorian metrology and its instrumentation. In R. Bud & S. Cozzens (eds), *Invisible Connections* (pp. 25–54). Bellingham: SPIE Optical Engineering Press.

Schama, S. (2004). *Citizens: A Chronicle of the French Revolution*. London: Penguin.

Scott, J.C. (1998). *Seeing Like a State: How Certain Schemes to Improve the Human Condition have Failed*. New Haven: Yale University Press.

Seager, J. (1993). *Earth Follies: Feminism, Politics and the Environment*. London: Earthscan.

Shapin, S. (1988). The house of experiment in seventeenth-century England. *Isis*, 79: 373–404.

Shapin, S. (1994). *A Social History of Truth: Civility and Science in Seventeenth-Century England*. Chicago: University of Chicago Press.

Shapin, S. (1995). Here and everywhere: sociology of scientific knowledge. *Annual Review of Sociology*, **21**: 289–321.

Shapin, S. (1996). *The Scientific Revolution*. Chicago: University of Chicago Press.

Shapin, S. (2001). How to be antiscientific. In A. Jay & H. Collins (eds), *The One Culture? A Conversation about Science* (pp. 99–115). Chicago: University of Chicago Press.

Shapin, S. & Schaffer, S. (1985). *The Leviathan and the Air-Pump: Hobbes, Boyle and the Experimental Life*. Princeton, NJ: Princeton University Press.

Shaw, W.N. & Owens, J.S. (1925). *The Smoke Problem of Great Cities*. London: Constable & Co. Ltd.

Sheail, J. (2002). *An Environmental History of Twentieth-Century Britain*. Basingstoke: Palgrave Macmillan.

Smith, N. (2004). *American Empire: Roosevelt's Geographer and the Prelude to Globalization*. Berkeley: University of California Press.

Soja, E.W. (2000). *Postmetropolis*. Oxford: Blackwell.

Thornes, J.E. (forthcoming) Cultural climatology and the representation of sky, atmosphere, weather and climate in selected art works of Constable, Monet and Eliasson. Geoforum.

Thorsheim, P. (2006). *Inventing Pollution: Coal, Smoke, and Culture in Britain since 1800*. Athens: Ohio University Press.

The Times (1881), Smoke Coal Abatement Exhibition. 23 December, p. 11.

The Times (1882). Ernest Hart (Chairman of Smoke Abatement Institute) letter to Editor of *The Times*, 28 October, p. 4.

The Times (1909). Smoke Abatement Exhibition. 2 March, p. 10.

The Times (1936a). The Domestic Hearth, 2 October, p. 15.

The Times (1936b) Air pollution measured – smoke abatement exhibition deaths due to fog. 2 October, p. 11.

The Times (1952a). Transport dislocated by three days of fog. 8 December, p. 8.

The Times (1952b). Chaos again in fog – London queue of 3,000. 9 December, p. 8.

The Times (1952c). Opera discontinued. 9 December, p. 8.

The Times (1952d). Inquiry into London fog suggested – 'abnormal air pollution'. 20 December, p. 3.

The Times (1953a). Hospital admissions during fog – pressure on emergency bed service. 1 January, p. 3.

The Times (1953b). Effect of fog on 1952 death rate – influenza epidemic compared. 21 December, p. 4.

The Times (1953c). London air as public sewer – pollution investigation. 21 January, p. 3.

Urry, J. (2002). *The Tourist Gaze*, second edition. London: Sage.

United Nations Economic Commission for Europe (1975). Conference on Security and Cooperation in Europe – Final Act (Helsinki).

United Nations Economic Commission for Europe (2008). http://www.unece.org/welcome.html (accessed 15 July 2008).

Walkerdine, V. & Lucy, H. (1989). *Democracy in the Kitchen: Regulating Mothers and Socialising Daughters*. London: Virago.

Warren Spring Laboratory (1972a). *National Survey of Air Pollution – 1961–1971: Volume 1, Introduction, United Kingdom, South East and Greater London*. London: HMSO.

Warren Spring Laboratory (1972b). *National Air Pollution Survey – 1961–1971: Volume 5, Scotland, Northern Ireland, Accuracy of Data, Index*. London: HMSO.

Weale, A. (1992). *The New Politics of Pollution*. Manchester: Manchester University Press.

Whitehead, M. (in press). Domesticating technological myth: gender exhibition spaces and the clean air movement in the UK. *Social and Cultural Geography*.

Whitehead, M., Jones, R. & Jones, M. (2006). *The Nature of the State: Excavating the Political Ecologies of the Modern State*. Oxford: Oxford University Press.

Williams, A.J. (2007). Hakumat al Tayarrat: the role of air power in the enforcement of Iraq's boundaries. *Geopolitics*, **12**: 505–28.

Withers, C. (2001). *Geography, Science and National Identity.* Cambridge: Cambridge University Press.

World Health Organisation (2005). *WHO Air Quality Guidelines for Particulate Matter, Ozone, Nitrogen Dioxide and Sulphur Dioxide: Global Update, Summary of Risk Assessments.* Geneva: World Health Organisation.

Worster, D. (1994). *Nature's Economy: A History of Ecological Idea*, 2nd edition. Cambridge: Cambridge University Press.

Yanni, C. (1999). *Nature's Museums: Victorian Science and the Architecture of Display.* London: Athlone Press.

Index

Note: Page references in **bold** print refer to tables and those in *italics* refer to figures.